THE GREAT UNKNOWN

ALSO BY MARCUS DU SAUTOY

The Music of the Primes:

Why an Unsolved Problem in Mathematics Matters

The Numbers Mystery:

A Mathematical Odyssey through Everyday Life

Symmetry: A Journey into the Patterns of Nature

THE GREAT UNKNOWN

SEVEN JOURNEYS
TO THE FRONTIERS
OF SCIENCE

MARCUS
DU SAUTOY

VIKING

VIKING
An imprint of Penguin Random House LLC
375 Hudson Street
New York, New York 10014
penguin.com

First published in Great Britain as *What We Cannot Know*
by 4th Estate, an imprint of HarperColllinsPublishers

ISBN 9780735221802 (hardcover)
ISBN 9780735221819 (ebook)

Printed in the United States of America
1   3   5   7   9   10   8   6   4   2

Set in Minion Pro and Gotham
Designed by Amy Hill

To my parents,
who started me on my journey
to the edge of knowledge

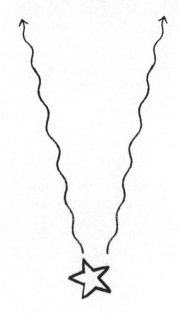

CONTENTS

THE GREAT UNKNOWN

ZERO

THE KNOWN UNKNOWNS

O

Everyone by nature desires to know.

—Aristotle, *Metaphysics*

E VERY WEEK, headlines announce new breakthroughs in our understanding of the universe, new technologies that will transform our environment, new medical advances that will extend our lives. Science is giving us unprecedented insights into some of the big questions that have challenged humanity ever since we've been able to formulate them. Where did we come from? What is the ultimate destiny of the universe? What are the building blocks of the physical world? How does a collection of cells become conscious?

In the last ten years alone we've landed a spaceship on a comet, built robots that can create their own language, used stem cells to repair the pancreas of diabetic patients, discovered how to use the power of thought to manipulate a robotic arm, and sequenced the DNA of a 50,000-year-old cave girl. Science magazines are bursting with the latest breakthroughs emerging from the world's laboratories. We know so much.

Science is our best weapon in our fight against fate. Instead of giving in to the ravages of disease and natural disaster, we have created vaccines to combat deadly viruses like polio and Ebola. As the world's population continues to escalate, scientific advances provide the best hope of feeding the 9.6 billion people who are projected to be alive in 2050. Science warns us about the deadly impact we are having on our environment and gives us the chance to do something about it before it is too late. An asteroid might have wiped out the dinosaurs, but science is our best shield against any future direct hits. In the human race's constant battle with death, science is its best ally.

Science is king not only when it comes to our fight for survival but also in improving our quality of life. We are able to communicate with friends and family across vast distances. We have created virtual worlds to which we can escape in our leisure time and we can re-create in our living rooms the great performances of Mozart, Miles, and Metallica at the press of a button.

The desire to know is programmed into the human psyche. Early humans with a thirst for knowledge were the ones who survived to transform their environment. Those not driven by that craving were left behind. Evolution has favored the mind that wants to know the secrets of how the universe works. The adrenaline rush that accompanies the discovery of new knowledge is nature's way of telling us that the desire to know is as important as the drive to reproduce. As Aristotle suggested in the opening line of *Metaphysics*, understanding how the world works is a basic human need.

When I was a schoolkid, science very quickly captivated me. I fell in love with its extraordinary power to reveal the workings of the universe. The fantastic stories that my science teachers told me seemed even more fanciful than the fiction I'd been reading at home. I persuaded my parents to buy me a subscription to *New Scientist* and devoured *Scientific American* in our local library. I hogged the television each week to watch episodes of *Horizon* and *Tomorrow's World*. I was enthralled by Jacob Bronowski's *Ascent of Man*, Carl Sagan's *Cosmos*, and Jonathan Miller's *Body in Question*. Every Christmas, the Royal Institution Christmas Lectures provided a dollop of science alongside our family turkey. My stocking was stuffed with books by George Gamow and Richard Feynman. It was a heady time, with new break-throughs announced each week.

Alongside these stories of discovery, I began to get fired up by the untold tales. What we knew lay in the past but we didn't yet know the future, my future. I became obsessed with the puzzle books of Martin Gardner that my math teacher gave me. The excitement of wrestling with a conundrum and the sudden release of euphoria as I cracked each puzzle got me addicted to the drug of discovery. Those puzzles were my training ground for the greater challenge of tackling questions that didn't have an answer in the back of the book. It was the unanswered questions, the mathematical mysteries and scientific puzzles that no one had cracked, that would become the fuel for my life as a scientist.

It is quite extraordinary how much more we have understood about the universe even in the half century that I've been alive. Technology has extended our senses so we can see things that were beyond the conception of the scientists who excited me as a kid. A new range of telescopes that look out at the night sky enabled us to discover planets like Earth that could be home to intelligent life. They have revealed the amazing fact that three quarters of the way into the lifetime of our universe, its expansion started to accelerate. I remember reading as a kid that we were in for a big crunch, but now it seems that we have a completely different future awaiting us.

Particle colliders like the Large Hadron Collider at CERN have allowed us to penetrate the inner workings of matter itself, revealing new particles—like the top quark discovered in 1994 and the Higgs boson discovered in 2012—that were bits of speculative mathematics when I was reading my *New Scientist* at school. And since the early '90s the fMRI scanner has allowed us to look inside the brain and discover things that were not even considered part of the remit of science when I was a kid back in the '70s. The brain was the preserve of philosophers and theologians, but today technology can reveal when you are thinking about Jennifer Aniston or predict what you are going to do next even before you know it yourself.

Biology has seen an explosion of breakthroughs. In 2003 it was announced that scientists had mapped an entire human DNA sequence consisting of 3 billion letters of genetic code. In 2011 the complete neuronal network of the *C. elegans* worm was published, providing a complete picture of how the 302 neurons in the worm are connected. Chemists, too, have been breaking new territory. A totally new form of carbon was discovered in 1985, which binds together like a football; and chemists surprised us again in 2003 by creating the first examples of graphene, showing how carbon can form a honeycomb lattice one atom thick.

In my lifetime the subject to which I would eventually dedicate myself, mathematics, has seen some of the great enigmas finally resolved: Fermat's Last Theorem and the Poincaré conjecture, two challenges that had outfoxed generations of mathematicians. New mathematical tools and insights have opened up hidden pathways to navigate the mathematical universe. Keeping up with all these new advances, let alone making your own contribution, is a challenge in its own right.

A few years ago I got a new job title to add to my role as a professor of mathematics at Oxford: the Simonyi Professor for the Public Understanding of Science. There seems to be a belief that with such a title I should know it all. People ring me up expecting me to know the answer to every scientific question. Shortly after I'd accepted the job, the Nobel Prize for medicine was announced. A journalist called, hoping for an explanation of the breakthrough that was being rewarded: the importance of telomeres.

Biology has never been my strong point, but I was sitting in front of my computer screen and so I'm embarrassed to admit I got the Wikipedia page up on telomeres and, after a quick scan, proceeded to explain authoritatively that they are the bit of genetic code at the end of our chromosomes that controls aging, among other things. The technology we have at our fingertips has increased that sense that we have the potential to know anything. Tap any question into a search engine and the device seems to predict, even before you finish typing, what you want to know and provides a list of places to find the answer.

But understanding is different from a list of facts. Is it possible for any scientist to know it all? To know how to solve nonlinear partial differential equations? To know how SU(3) governs the connection between fundamental particles? To know how cosmological inflation gives rise to the state of the universe? To know how to solve Einstein's equations of general relativity or Schrödinger's wave equation? To know how neurons and synapses trigger thought? Newton, Leibniz, and Galileo were perhaps the last scientists to know all that was known.

I must admit that the arrogance of youth infused me with the belief that I could understand anything that was known. With enough time, I thought, I could crack the mysteries of mathematics and the universe, or at least master the current lay of the land. But increasingly, I am beginning to question that belief, to worry that some things will forever remain beyond my reach. Often my brain struggles to navigate the science we currently know. Time is running out to know it all.

My own mathematical research is already pushing the limits of what my human brain feels capable of understanding. I have been working for more than ten years on a conjecture that remains stubbornly resistant to my attempts to crack it. My new role as the Professor for the Public Understanding of Science has pushed me outside the comfort zone of mathematics into the messy concepts of neuroscience, the slippery ideas of philosophy, the unfounded theories of physics. It has required a way of thinking that is alien to my mathematical mode of thought, which deals in certainties, proofs, and precision. My attempts to understand everything currently regarded as scientific knowledge have severely tested the limits of my own ability to understand.

We stand on the shoulders of giants, as Newton famously declared. And so my own journey to the frontiers of knowledge has pushed me to explore how others have articulated their work, to listen to lectures and seminars by those immersed in the fields I'm trying to understand, and to talk to those pushing the boundaries of what is known, questioning contradictory stories and consulting the evidence recorded in scientific journals. How much can you trust any of these stories? Just because the scientific community accepts a story as the current best fit doesn't mean it is true. Time and again, history reveals the opposite to be the case, and this must always act as a warning that current scientific knowledge is provisional. Mathematics has a slightly different quality, as a proof provides the chance to establish a more permanent state of knowledge. But even when I am creating a new proof, I will often quote results by fellow mathematicians whose proofs I haven't checked myself. To do so would mean running in order to keep still.

For any scientist the real challenge is not to stay within the secure garden of the known but to venture out into the wilds of the unknown. That is the challenge at the heart of this book.

WHAT WE DON'T KNOW

Despite all the breakthroughs made over the last centuries, there are still lots of deep mysteries waiting out there for us to solve. Things we don't know. The knowledge of what we don't know seems to expand faster than our catalog of breakthroughs. The known unknowns outstrip the known knowns. And it is those unknowns that drive science. A scientist is more interested in the things he or she can't understand than in telling all the stories we already know the answers to. Science is a living, breathing subject because of all those questions we can't answer.

For example, the stuff that makes up the physical universe we interact with seems to account for only 4.9 percent of the total matter content of our universe. So what is the other 95.1 percent of so-called dark matter and dark energy made up of? If our universe's expansion is accelerating, where is all the energy coming from that fuels that acceleration?

Is our universe infinite? Are there infinitely many other infinite universes parallel to our own? If there are, do they have different laws of physics? Were there other universes before our universe emerged from the Big Bang? Did time exist before the Big Bang? Does time exist at all, or does it emerge as a consequence of more fundamental concepts?

How can we unify Einstein's theory of general relativity, the physics of the very large, with quantum physics, the physics of the very small? This is the search for something called quantum gravity, an absolute necessity if we are ever going to understand the Big Bang.

And what of the understanding of our human body, something so complex that it makes quantum physics look like a high school exercise? We are still trying to come to grips with the complex interaction between gene expression and our environment. Can we find a cure for cancer? Is it possible to beat aging? Could there be someone alive today who will live to be a thousand years old?

And what about where humans came from? Evolution is a process of random mutations, so would a different roll of the evolutionary dice still produce organisms with eyes? If we rewound evolution and pressed "play," would we still get intelligent life, or are we the result of a lucky roll of the dice? Is there intelligent life elsewhere in our universe? And what of the technology we are creating? Can a computer ever attain consciousness? Will I eventually be able to download my consciousness so that my mind can survive the death of my body?

Mathematics, too, is far from finished. Despite popular belief, Fermat's Last Theorem was not the last theorem. Mathematical unknowns abound. Are there any patterns in prime numbers, or are they outwardly random? Will we be able to solve the mathematical equations for turbulence? Will we ever understand how to factorize large numbers efficiently?

Despite so much that is still unknown, scientists are optimistic that these questions won't remain unanswered forever. The last few decades give us reason to believe that we are in a golden age of science. The rate of discoveries in science appears to grow exponentially. In 2014 the science journal *Nature* reported that the number of scientific papers has been doubling every nine years since the end of World War II. Computers are also developing at an extraordinary rate. Moore's Law has it that computer processing power will double every two years. Ray Kurzweil believes that the same applies to technological progress: that the rate of change over the next hundred years will be comparable to what we've experienced in the last 20,000 years.

Can scientific discovery really sustain this growth? Kurzweil talks about the Singularity, a moment when the intelligence of our technology will exceed human intelligence. Is scientific progress destined for its own singularity, a moment when we know it all? Surely at some point we might actually discover the underlying equations that explain how the universe works. We will discover the final particles that make up the building blocks of the physical universe and how they interact with each other. Some scientists believe that the current rate of scientific progress will lead to a moment when we might discover a theory of everything. They even give it a name: ToE.

As Stephen Hawking declared in *A Brief History of Time*, "I believe there are grounds for cautious optimism that we may be near the end of the search for the ultimate laws of nature." He concludes dramatically with the provocative statement that then "we would know the mind of God."

Is such a thing possible? To know everything? Would we want to know everything? Scientists have a strangely ambivalent relationship with the unknown. On the one hand, what we don't know is what intrigues and fascinates us, and yet the mark of success as a scientist is resolution and knowledge, to make the unknown known.

Are there limits to what we can discover? Are there quests that will never be resolved? Are some regions beyond the predictive powers of science and mathematics—like time before the Big Bang? Are there ideas so complex that they exceed the conception of our finite human brains? Can brains really investigate themselves, or does the analysis enter an infinite loop from which it is impossible to rescue itself? Are there mathematical conjectures that can never be proved true?

It seems defeatist, even dangerous, to acknowledge such questions. While the unknown is the driving force for doing science, the unknowable is science's nemesis. As a fully signed-up member of the scientific community, I hope that we can ultimately answer the big open questions. So it seems important to know whether the expedition I've joined will hit boundaries beyond which we cannot proceed. Are there in fact any questions that won't ever get closure?

That is the challenge I've set myself in this book. I want to know whether there are things that, by their very nature, we will never know. Are there things that will always be beyond the limits of knowledge? Despite the marauding pace of scientific advances, are there things that will remain beyond the reach of even the greatest scientists? Mysteries that will forever remain part of the great unknown?

It is, of course, very risky at any point in history to try to articulate the Things We Cannot Know. How can you know what new insights will suddenly pull the unknown into the knowable? This is partly why it is useful to look at the history of how we came to know the things we know, because it reveals how often we've been at a point where we think we have reached the frontier, only to find a greater landscape beyond.

Take the statement made by French philosopher Auguste Comte in 1835 about the stars: "We shall never be able to study, by any method, their chemical composition or their mineralogical structure." An absolutely fair statement given that this knowledge seemed to depend on our visiting the star. What Comte hadn't considered was the possibility that the star could visit us, or at least that photons of light emitted by the star could reveal its chemical makeup.

A few decades after Comte's prophecy, scientists had determined the chemical composition of our own star, the sun, by analyzing the spectrum of light emitted. As the nineteenth-century British astronomer Warren de la Rue declared: "If we were to go to the Sun, and to bring some portions of it and analyze them in our laboratories, we could not examine them more accurately than we can by this new mode of spectrum analysis."

Scientists went on to determine the chemical composition of stars we are unlikely ever to visit. As science in the nineteenth century continued to give us an ever greater understanding of the mysteries of the universe, there began to emerge a feeling that we might eventually have a complete picture.

In 1900 Lord Kelvin, regarded by many as one of the greatest scientists of his age, believed that moment had come. He declared to the meeting of the British Association of Science: "There is nothing new to be discovered in physics now. All that remains is more and more precise measurement." American physicist Albert Abraham Michelson concurred. He too thought that the future of science would simply consist of adding a few decimal places to the results already obtained. "The more important fundamental laws and facts of physical science have all been discovered . . . our future discoveries must be looked for in the sixth place of decimals."

Five years later, Einstein announced his extraordinary new conception of time and space, followed shortly by the revelations of quantum physics. Kelvin and Michelson couldn't have been more wrong.

What I want to try to explore is whether there are problems that we can prove will forever remain beyond our knowledge. Perhaps there are none. As a scientist, that is my hope. One of the dangers when faced with currently unanswerable problems is to give in too early to their unknowability. But if there are unanswerables, what status do they have? Can you choose from the possible answers and it won't really matter which one you opt for?

Talk of known unknowns is not reserved to the world of science. Secretary of Defense Donald Rumsfeld strayed into the philosophy of knowledge with the famous declaration:

> There are known knowns; there are things that we know that we know. We also know there are known unknowns; that is to say, we know there are some things we do not know. But there are also unknown unknowns, the ones we don't know we don't know.

Rumsfeld received a lot of flack for this cryptic response to a question fired at him during a briefing at the Department of Defense about the lack of evidence connecting the government of Iraq with weapons of mass destruction. Journalists and bloggers had a field day, culminating in Rumsfeld being given the Foot in Mouth award by the Plain English Campaign. And yet if one unpacks the statement, Rumsfeld very concisely summed up different types of knowledge. He perhaps missed one interesting category: the unknown knowns, the things that you know yet dare not admit to knowing. As the philosopher Slavoj Zizek argues, these are possibly the most dangerous, especially when held by those with political power. This is the domain of delusion. Repressed thoughts. The Freudian unconscious.

I would love to tell you about the unknown unknowns, but then they'd be known! Nassim Taleb, author of *The Black Swan*, believes that the emergence of unknowns is responsible for the biggest changes in society. For Kelvin, relativity and quantum physics turned out to be the great unknown unknown that he was unable to imagine. My hope in this book is to articulate the known unknowns and ask whether any will remain forever unknown.

I have called these unknowns "Edges." There are seven of them, and each one represents the horizon beyond which we cannot see. My journey to the Seven Edges of knowledge will pass through the known knowns, to demonstrate how we have traveled beyond what we previously thought were the limits of knowledge. This journey will also test my own ability to grasp what is known, because it's becoming increasingly challenging as a scientist to know even the knowns.

As much as this book is about what we cannot know, it is also important to understand what we do know and how we know it. My journey to the frontiers of knowledge will take me through the terrain that scientists have already mapped, to the very limits of today's breakthroughs. On the way I will stop to consider those moments when scientists thought they had hit a wall beyond which progress was no longer possible, only for the next generation to find a way. This will give us an important perspective on those problems that we might think are unknowable today. By the end of our journey, I hope this book will provide a comprehensive survey not just of what we cannot know but also of the things we do know.

To help me through these areas of science that are outside my comfort zone, I have enlisted the help of experts to guide me as I reach each science's Edge and to test whether it is my own limitations or limitations inherent in the questions I am tackling that make these questions unknowable.

What happens then if we encounter a question that cannot be answered? How does one cope with not knowing? Dare I admit to myself that some things will forever remain beyond my reach? How do we cope with not knowing? That challenge has elicited some interesting responses from humans across the millennia, not least the creation of an idea called God.

TRANSCENDENCE

There is another reason why I have been driven to investigate the unknowable, which is also related to my new job. The previous incumbent of the chair for the Public Understanding of Science was a certain Richard Dawkins. When I took over the position from Dawkins I braced myself for the onslaught of questions that I would get, not about science, but about religion. The publication of *The God Delusion* and his feisty debates with creationists resulted in Dawkins spending the later years of his tenure debating questions of religion and God.

So it was inevitable that when I took up the chair people would be interested in my stance on religion. My initial reaction was to distance myself from the debate about God. My job was to promote scientific progress and to engage the public in the breakthroughs happening around them. I was keen to move the debate back to questions of science rather than religion.

In an urban environment like London, football has taken over the role that religion played in society of binding a community together, providing rituals that they can share. For me, the science that I began to learn as a teenager did a pretty good job of pushing out any vaguely religious thoughts I had as a kid. I sang in my local church choir, which exposed me to the ideas that Christianity had to offer for understanding the universe. School education in the 1970s in the United Kingdom was infused with mildly religious overtones: renditions of "All Things Bright and Beautiful" and the Lord's Prayer in assemblies. Religion was dished up as something too simplistic to survive the sophisticated and powerful stories that I would learn in the science labs at my secondary school. Religion was quickly pushed out. Science . . . and football . . . were much more attractive.

Inevitably the questions about my stance on religion would not be fobbed off with such a flippant answer. I remember that during one radio interview on a Sunday morning on BBC Northern Ireland I was gradually sucked into considering the question of the existence of God. I guess I should have seen the warning signs. On a Sunday morning in Northern Ireland, God isn't far from the minds of many listeners.

As a mathematician I am often faced with the challenge of proving the existence of new structures or coming up with arguments to show why such structures cannot exist. The power of the mathematical language to produce logical arguments has led a number of philosophers throughout the ages to resort to mathematics as a way of proving the existence of God. But I always have a problem with such an approach. If you are going to prove existence or otherwise in mathematics, you need a very clear definition of what it is that you are trying to prove exists.

So after some badgering by the interviewer about my stance on the existence of God, I pushed him to try to define what God meant for him so that I could engage my mathematical mind. "It is something which transcends human understanding." At first I thought: what a cop-out. You have just defined it as something that by its very nature I can't get a handle on. But I became intrigued by this definition. Perhaps it wasn't such a cop-out after all.

What if you *define* God as the things we cannot know. The gods in many ancient cultures were often placeholders for the things people couldn't explain or understand. Our ancestors found volcanic eruptions or eclipses so mysterious that they became acts of gods. As science has explained such phenomena, these gods have retreated.

This definition has some things in common with a God commonly called the "God of the gaps." This phrase was generally used as a derogatory term by religious thinkers who could see that this God was shrinking in the face of the onslaught of scientific knowledge, and a call went out to reject this kind of God. The phrase "God of the gaps" was coined by the Oxford mathematician and Methodist church leader Charles Coulson, when he declared: "There is no 'God of the gaps' to take over at those strategic places where science fails."

But the phrase is also associated with a fallacious argument for the existence of God, one that Richard Dawkins spends some time shooting down in *The God Delusion*: if there are things that we can't explain or know, there must be a God at work filling the gap. I am more interested not in the existence of a God to fill the gap, but in equating God with the abstract idea of the things we cannot know. Not in the things we currently don't know, but the things that by their nature we can never know—the things that will always remain transcendent.

Religion is more complex than the simple stereotype often offered up by modern society. For many ancient cultures in India, China, and the Middle East, religion was not about worshiping a supernatural intelligence so much as it was an attempt to appreciate the limits of our understanding and of language. As the theologian Herbert McCabe declared, "To assert the existence of God is to claim that there is an unanswered question about the universe." Science has pushed hard at those limits. So is there anything left? Will anything always be beyond the limit? Does McCabe's God exist?

This is the quest at the heart of this book. But first we need to know if, in fact, anything will remain unanswered about the universe. Is there really anything we cannot know?

FIRST EDGE

CHAOS

1

The unpredictable and the predetermined unfold together to make everything the way it is. It's how nature creates itself, on every scale, the snowflake and the snowstorm. It makes me so happy. To be at the beginning again, knowing almost nothing.

—Tom Stoppard, *Arcadia*

A SINGLE RED DIE sits next to me on my desk. I got it on a trip to Las Vegas. I fell in love with it when I saw it on the craps table. It was so perfectly engineered: such precise edges coming to a point, the face so smooth you couldn't feel what number you were touching. The pips are carved out and then filled in with paint that has the same density as the plastic used to make the die itself. This ensures that the face representing the 6 isn't a touch lighter than the opposite side, with a single pip. The feeling of the die in my hand is incredibly satisfying. It is a thing of beauty.

And yet I hate it.

Three pips are pointing up at me at the moment. But if I pick it up and let it fall from my hand I have no way of knowing how it will land. Dice are the ultimate symbol of the unknowable. The future seems knowable only when it becomes the past.

I have always been extremely unsettled by things that I cannot work out. I don't mind not knowing something, provided there is some way ultimately to calculate the answer—with enough time. Is the fate of this perfect Las Vegas die truly unknowable? Or with enough information can I actually deduce its next move? Surely it's just a matter of applying the right laws of physics and solving the appropriate mathematical equations. Surely this is something I can figure out. Or is it?

My subject, mathematics, was invented to give people a glimpse of what's out there, to look into the future—to become masters of fate, not its servants. Mathematics is the science of patterns. Being able to spot a pattern is a powerful tool in the evolutionary fight for survival. The pattern of the sun means that I can rely on its rising in the sky tomorrow or on the moon running through twenty-eight sunrises before it becomes full again. The caves in Lascaux show how counting thirteen quarters of the moon from the first winter rising of the Pleiades will bring you to a time in the year when the horses are pregnant and easy to hunt. Being able to predict the future is the key to survival.

But some things appear to have no pattern or appear to have patterns so complex that they are beyond our ability to spot them. An individual roll of the dice is not like the rising of the sun. There seems to be no way of knowing which of the six faces will be pointing upward once the die finally comes to rest. This is why dice have been used since antiquity as a way to decide disputes, to play games, to wager money.

On a recent trip to Israel I took my children to an archeological dig at Beit Guvrin. It was such a popular settlement in ancient times that the site consists of layer upon layer of cities built on top of one another. There is so much stuff in the ground that the archeologists are happy to enlist amateurs like me and my kids to help excavate the site, even if a few pots are broken along the way. Sure enough, we pulled out lots of pottery shards, but we also kept unearthing animal bones. We thought they were the remains of dinner, but our guide explained that in fact they were the earliest form of dice.

Archeological digs of settlements dating back to Neolithic times have revealed a disproportionately high density of heel bones of sheep and other animals among the shattered pottery and flints that are usually found in such sites. These bones are, in fact, the ancestors of my casino dice. When thrown, the bones naturally land on one of four sides. Often there are letters or numbers carved into each side. These early dice are thought to have been used for divination, connecting the outcome of the roll of the dice to the will of the gods. Knowledge of how the dice would land was believed to transcend human understanding.

Over time, dice assumed a more prosaic place as part of our world of leisure. The first cube-shaped dice like the one on my desk were found around Harappa in what is now northeast Pakistan, where one of the first urban civilizations evolved, dating back to the third millennium BC. At the same time, you find four-faced pyramid dice appearing in a game that was discovered in the city of Ur, in ancient Mesopotamia. The Romans and Greeks were addicts of games of dice, as were medieval soldiers, who returned from the Crusades with a new game called hazard, derived from the Arabic word for dice: *al-zahr*. It was an early version of craps, the game that is being played in the casinos in Vegas.

If I could predict the roll of the dice, the many games that depend on them would never have caught on. The excitement of backgammon or craps comes from not knowing what number you will throw. So perhaps gamers won't thank me as I try to break the mystery and predict the roll of my dice.

For centuries no one even thought that such a feat was possible. The ancient Greeks, who were among the first to develop mathematics as a tool to navigate their environment, didn't have any clue as to how to tackle such a dynamic problem. Their mathematics was a static, rigid world of geometry, not one that could cope with things tumbling across the floor. They could produce formulas to describe the contours of a cube, but once the cube started moving they were lost.

Aristotle believed that events could essentially be classified into three categories: "certain events" that happen by necessity following the laws of nature; "probable events" that happen in most cases but could have a few exceptions; and finally "unknowable events" that happened by pure chance. Aristotle put the roll of dice firmly in the last category.

As Christian theology made its impact on philosophy, matters worsened. Since the roll of the dice was in the hands of God, it was not something that humans could aspire to know. As St. Augustine put it, "We say that those causes that are said to be by chance are not non-existent but are hidden, and we attribute them to the will of the true God."

There was no such thing as chance, no free will. The unknowable was known to God, who determined the outcome. Any attempt to predict the roll of the dice was the work of a heretic, someone who dared to think they could know the mind of God. King Louis XI of France even went as far as to prohibit the manufacture of dice, believing that games of chance were ungodly. It wasn't until the sixteenth century that dice were wrestled out of the hands of God and put into the hands, and minds, of humans.

I've put two more dice next to my beautiful Las Vegas die. So here's a question: If I throw all three dice, is it better to bet on a score of 9 or a score of 10 coming up? Prior to the sixteenth century no tools were available to answer such a simple question. And yet anyone who had played for long enough would know that if I were throwing only two dice then it would be wise to bet on 9 rather than 10. Experience would tell you before too long that, on average, you get 9 a third more often than you get 10. With three dice it is harder to get a feel for which way to bet, because 9 and 10 seem to occur equally often. But is that really true?

In Italy at the beginning of the sixteenth century, an inveterate gambler by the name of Girolamo Cardano first realized that there are patterns that can be exploited in a game of dice. They weren't patterns that could be used on an individual throw. Rather, they emerged over the long run, patterns that a gambler like Cardano, who spent many hours throwing dice, could use to his advantage. So addicted was he to gambling that on one occasion he even sold his wife's possessions to raise the funds for the table stakes.

Cardano had the clever idea of counting how many different futures the dice could have. If I throw two dice, there are thirty-six different futures. They are depicted in the following diagram.

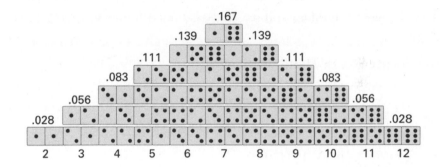

Only three of them total 10, while four give you a score of 9. So Cardano reasoned that if you are throwing two dice, it makes sense to bet on 9 rather than 10. It did not help in any individual game, but in the long run it meant that Cardano would come out on top. Unfortunately, while he was a disciplined mathematician, he wasn't very disciplined when it came to gambling. He managed to lose all of his father's inheritance and would regularly get into knife fights with his opponents when the dice went against him.

He was nevertheless determined to get one prophecy correct. He had predicted the date of his death: September 21, 1576. To make sure he got this bet right he took matters into his own hands and committed suicide when the date finally struck. As much as I crave knowledge, I think this is going a little far. Indeed, knowing the date of your death is something that most people would prefer to opt out of. But Cardano was determined to win, even when he was dicing with Death.

Before taking his life, he wrote what many consider to be the first book that made inroads into predicting the behavior of dice as they roll across the table. Although written around 1564, *Liber de Ludo Aleae* didn't see the light of day until it was eventually published in 1663.

The great Italian physicist Galileo Galilei applied the same analysis as Cardano to decide whether to bet on a score of 9 or 10 when three dice are thrown. He reasoned that there were $6 \times 6 \times 6 = 216$ different futures the dice could take. Of these, twenty-five gave you a 9 while twenty-seven gave you a 10. Not a big difference, and one that would be difficult to pick up, but large enough so that betting on 10 should give you an edge in the long run.

PASCAL'S WAGER

The mathematical mastery of dice shifted from Italy to France in the mid-seventeenth century, when two big hitters, Blaise Pascal and Pierre de Fermat, started applying their minds to predicting the future of these tumbling cubes. Pascal had become interested in trying to understand the outcome of the roll of the dice after meeting one of the great gamblers of the day, the Chevalier de Méré. De Méré had challenged Pascal with a number of interesting scenarios. One was the problem Galileo had cracked. Others included whether it was advisable to bet that at least one 6 will appear if a die is thrown four times, and a popular challenge at the time called the "problem of points."

Pascal entered into a lively correspondence with the great mathematician and lawyer Pierre de Fermat and they tried to sort out the problems set forward by de Méré. With the throw of four dice, one could consider the $6 \times 6 \times 6 \times 6 = 1,296$ different outcomes and count how many include a 6, but that becomes pretty cumbersome. Instead, Pascal reasoned that there is a 5/6 chance that you won't see a 6 with one throw. Since each throw is independent, that means there is a $5/6 \times 5/6 \times 5/6 \times 5/6 = 625/1{,}296 = 48.2$ percent chance that you won't get a 6 in four throws. Which means there is a 51.8 percent chance that you will see a 6—just above an even chance, so worth betting on.

The problem of "points" was even more challenging. Suppose two players—let's call them Fermat and Pascal—are rolling a single die. Fermat scores a point if it lands on 4 or higher; Pascal scores a point otherwise. Each, therefore, has a 50:50 chance of winning a point on any roll of the die. They've wagered sixty-four francs, which will go to the first to score three points. The game is interrupted, however, and can't be continued, when Fermat is on two points and Pascal is on one point. How should they divide the sixty-four francs?

Traditional attempts to solve the problem focused on what had happened in the past. Maybe, having won twice as many rounds as Pascal, Fermat should get twice the winnings. This would make no sense if Fermat had won only one round before the game was interrupted. Pascal would get nothing but still had a chance of winning. Niccolò Fontana Tartaglia, a contemporary of Cardano, believed, after much thought, that this conundrum had no solution: "The resolution of the question is judicial rather than mathematical, so that in whatever way the division is made there will be cause for litigation."

Others weren't so certain. Attention turned not to the past, but to what could happen in the future. Their goal was not to predict the roll of the die but to imagine all the different future scenarios and divide the spoils according to which version of the future favored which player.

It is easy to get fooled here. There seem to be three scenarios. Fermat wins the next round and pockets sixty-four francs. Pascal wins the next round, resulting in a final round that either man wins. Fermat wins in two out of these three scenarios, so perhaps he should get two thirds of the winnings. This was the trap that de Méré fell into. Pascal argued that this wasn't correct: "The Chevalier de Méré is very talented but he is not a mathematician; this is, as you know, a great fault." A great fault, indeed!

Pascal argued that the spoils should be divided differently. There was a 50:50 chance that Fermat would win in one round, in which case he would get sixty-four francs. But if Pascal won the next round, then the two friends were equally likely to win the final round, so could divide the spoils thirty-two francs each. In either case, Fermat is guaranteed thirty-two francs. So the other thirty-two francs should be split equally, giving Fermat forty-eight francs in total.

Fermat, writing from his home near Toulouse, concurred with Pascal's analysis: "You can now see that the truth is the same in Toulouse as in Paris."

Pascal and Fermat's analysis of the game of points could be applied to much more complex scenarios. Pascal discovered that the secret to deciding the division of the spoils is hidden inside something now known as Pascal's triangle.

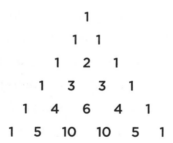

The triangle is constructed in such a way that each number is the sum of the two numbers immediately above it. The numbers you get are key to dividing the spoils in any interrupted game of points. For example, if Fermat needs two points for a win while Pascal needs four, then you consult the 2 + 4 = 6th row of the triangle and add the first four numbers together and the last two. This is the proportion in which you should divide the spoils. In this case it's a 1 + 5 + 10 + 10 = 26 to 1 + 5 = 6 division. So Fermat gets 26/32 × 64 = ₣52 and Pascal gets $6/32 \times 64 = ₣12$. In general, a game where Fermat needs n points to Pascal's m points can be decided by consulting the $(n + m)$th row of Pascal's triangle.

The French may have been beaten by several millennia to the discovery that this triangle is connected to the outcome of games of chance. The Chinese were inveterate users of dice and other methods like the *I Ching* to try to predict the future. The text of the *I Ching* dates back some three thousand years and contains precisely the same table that Pascal produced to analyze the outcomes of tossing coins, but today the triangle is attributed to Pascal rather than the Chinese.

Pascal wasn't interested only in dice. He famously applied his new mathematics of probability to one of the great unknowns: the existence of God.

"God is, or He is not." But to which side shall we incline? Reason can decide nothing here. There is an infinite chaos which separated us. A game is being played at the extremity of this infinite distance where heads or tails will turn up. . . . Which will you choose then? Let us see. Since you must choose, let us see which interests you least. You have two things to lose, the true and the good; and two things to stake, your reason and your will, your knowledge and your happiness; and your nature has two things to shun, error and misery. Your reason is no more shocked in choosing one rather than the other, since you must of necessity choose. . . . But your happiness? Let us weigh the gain and the loss in wagering that God is. . . . If you gain, you gain all; if you lose, you lose nothing. Wager, then, without hesitation that He is.

Called Pascal's wager, the argument is hardly compelling. It hinges on the belief that the payout would be much greater if one opted for a belief in God. You lose little if you are wrong and win eternal life if correct. On the other hand, wager against the existence of God and losing results in eternal damnation, while winning gains you nothing beyond the knowledge that there is no God. The argument falls to pieces if the probability of God existing is actually zero. Even if it isn't, the cost of belief might be too high when set against the probability of God's existence.

The probabilistic techniques developed for dealing with uncertainty by mathematicians like Fermat and Pascal were incredibly powerful. Phenomena that were regarded as the expression of the gods were beginning to be within reach of the minds of men. Today these probabilistic methods are our best weapon for trying to navigate everything from the behavior of particles in a gas to the ups and downs of the stock market. Indeed, the very nature of matter itself seems to be at the mercy of the mathematics of probability, as we shall discover in the Third Edge, when we apply quantum physics to predict what fundamental particles are going to do when we observe them. But for someone searching for certainty, these probabilistic methods represent a frustrating compromise.

I certainly appreciate the great intellectual breakthrough that Fermat, Pascal, and others made, but it doesn't help me to know the outcome when I throw my dice. As much as I've studied the mathematics of probability, it has always left me with a feeling of dissatisfaction. The one thing any course on probability drums into you is that it doesn't matter how many times in a row you get a 6: this has no influence on what will happen on the next throw.

So is there some way of knowing how my dice will land? Or is that knowledge always going to be out of reach? Not according to the revelations of a scientist in England.

THE MATHEMATICS OF NATURE

Isaac Newton is my all-time hero in the fight against the unknowable. The idea that I could possibly know everything about the universe has its origins in Newton's revolutionary work *Philosophiae Naturalis Principia Mathematica*. First published in 1687, the book is dedicated to developing a new mathematical language that promised to unlock how the universe behaves. It was a dramatically new model of how to do science. The work "spread the light of mathematics on a science which up to then had remained in the darkness of conjectures and hypotheses," declared the French physicist Alexis Clairaut in 1747.

The *Principia Mathematica* is also an attempt to unify, to create a theory that describes the celestial and the earthly, the big and the small. Johannes Kepler had come up with laws that described the motions of the planets, laws he'd developed empirically by looking at data and trying to come up with equations that would explain the data. Galileo had described the trajectory of a ball flying through the air. It was Newton's genius to understand that these were two examples of a single phenomenon: gravity.

Born on Christmas Day in 1643 in the Lincolnshire town of Woolsthorpe, Newton was always trying to tame the physical world. He made clocks and sundials, constructed miniature mills powered by mice, sketched countless plans for buildings and ships, and drew elaborate illustrations of animals. The family cat disappeared one day, carried away by a hot-air balloon that Newton had made. His school reports, however, did not anticipate a great future, describing him as "inattentive and idle."

Idleness is not necessarily such a bad trait in a mathematician. It can be a powerful incentive to look for some clever shortcut to solve a problem rather than relying on hard labor. But it's not generally a quality that teachers appreciate. Newton was doing so badly at school that his mother decided the whole thing was a waste of time and that he'd be better off learning how to manage the family farm in Woolsthorpe. Unfortunately, Newton was equally hopeless at managing the family estate, so he was sent back to school. Although probably apocryphal, it is said that Newton's sudden academic transformation coincided with a blow to the head that he received from the school bully. Whether true or not, his academic transformation saw him suddenly excelling at school, culminating in a move to study at Cambridge University.

When bubonic plague swept through England in 1665, Cambridge was closed as a precaution. Newton retreated to his family house in Woolsthorpe. Isolation is often an important ingredient for coming up with new ideas. Newton hid himself away in his room and thought: "Truth is the offspring of silence and meditation. I keep the subject constantly before me and wait 'til the first dawnings open slowly, by little and little, into a full and clear light."

In the isolation of Lincolnshire, Newton created a new mathematical language that could capture the problem of a world in flux; today we call it calculus. This tool would be key to our discovery of predicting how the universe will behave. It is this language that gives me the hope of one day gleaning how my casino dice might land.

Calculus tries to make sense of what at first looks like a meaningless sum: zero divided by zero. This is the sum that I must calculate if I want to know my die's instantaneous speed as it falls through the air. The speed is constantly increasing as gravity pulls it to the ground. So how can I calculate its speed at any given moment of time? For example, how fast is it falling after one second? Speed is the distance traveled divided by the time elapsed. So I could record the distance it drops in the next second and that would give me an average speed over that period. But I want to know its precise speed. I could record the distance traveled over a shorter period of time, say half a second or a quarter of a second. The smaller the interval of time, the more accurately I will be calculating its speed. Ultimately, to get the precise speed, I will need to take an interval of time that is infinitesimally small. But then I am faced with calculating zero divided by zero.

Newton's calculus made sense of this calculation. He understood how to calculate speed as the interval of time becomes infinitesimally small. It was a revolutionary new language that managed to capture a changing dynamic world. The geometry of the ancient Greeks was per-

Calculus: Making Sense of Zero Divided by Zero

Suppose that a car starts from a stationary position. When the stopwatch starts, the driver slams his foot on the accelerator. Suppose that we record that after *t* seconds the driver has covered *t* x *t* meters. How fast is the car going after ten seconds? We get an approximation of the speed by looking at how far the car has traveled in the period from ten to eleven seconds. The average speed during this time is (11 x 11 − 10 x 10)/1 = 21 meters per second.

But if we look at a smaller window of time, say the average speed over 0.5 seconds, we get:

(10.5 x 10.5 − 10 x 10)/0.5 = 20.5 meters per second.

Slightly slower, of course, because the car is accelerating, so on average it is going faster in the second half second. But what if we take an even smaller snapshot. What about halving the window of time again:

(10.25 x 10.25 − 10 x 10)/0.25 = 20.25 meters per second.

I hope the mathematician in you has spotted the pattern. If I take a window of time that is *x* seconds, the average speed over this time will be 20 + *x* meters per second. The speed as I take smaller and smaller windows of time is getting closer and closer to twenty meters per second. So, although to calculate the speed at 10 seconds looks like I have to figure out the calculation 0/0, the calculus makes sense of what this should mean.

fect for a static, frozen picture of the world. Newton's mathematical breakthrough could describe a moving world. Mathematics had gone from describing a still life to capturing a moving image. This was the scientific equivalent of how the dynamic art of the Baroque period burst forth from the static art of the Renaissance. Newton later looked back at this time as one of the most productive of his life, calling it his *annus mirabilis*: "I was in the prime of my age for invention and minded Mathematicks and Philosophy more than at any time since."

Everything around us is in a state of flux, so it was perhaps no wonder that his discovery would be so influential. But for Newton calculus was a means to an end, a personal tool that helped him reach the scientific conclusions that he documents in the *Principia*, the great treatise published in 1687 that describes his ideas on gravity and the laws of motion.

Writing in the third person, he explained that his calculus was key to the scientific discoveries contained inside: "By the help of this new Analysis Mr. Newton found out most of the propositions in the *Principia*." And yet no account of the "new analysis" is published. He privately circulated his ideas among friends, but he felt no urge to publish them for others to appreciate.

Fortunately calculus is now widely taught. It is a language that I spent years learning and gaining fluency in as a mathematical apprentice. To know my dice, I will need to mix Newton's mathematical breakthrough with his great contribution to physics: the famous laws of motion with which he opens his *Principia*.

A THEORY OF EVERYTHING

Newton outlines in the *Principia* three simple laws from which so much of the dynamics of the universe evolve.

Newton's First Law of Motion: *A body will continue in a state of rest or uniform motion in a straight line unless it is compelled to change that state by forces acting on it.*

This sounds obvious, but it was not so obvious to the likes of Aristotle. If you roll a ball along a flat surface it comes to a rest. It looks like you need a force to keep it moving. There is, however, a hidden force that is changing its speed: friction. If I throw my dice in outer space, away from any gravitational fields, then they will carry on flying in a straight line at constant speed ad infinitum.

In order to change an object's speed or direction you need a force to act against it. Newton's second law explained how that force would change its motion. Calculus has already allowed me to calculate the speed of my die as it falls down toward the table. The rate of change in that speed may be deduced by applying calculus again. Newton's second law holds that there is a direct relationship between the force being applied and the rate of change in speed.

Newton's Second Law of Motion: *The rate of change of motion, or acceleration, is proportional to the force that is acting on it and inversely proportional to its mass.*

To understand the speed of my cascading dice, I need to understand the forces acting on them. Newton's Universal Law of Gravitation identified one of the principal forces affecting everything from the fall of an apple to the movement of a planet through the solar system. The law states that the force acting on a body of mass m_1 by another body of mass m_2 that is a distance of r away is equal to

$$\frac{G \times m_1 \times m_2}{r^2}$$

where G is an empirical physical constant that controls how strong gravity is in our universe.

With these laws I can now describe the trajectory of a ball flying through the air, or a planet traveling through the solar system, or of dice falling from my hand. The next problem occurs when the dice hit the table. What happens then? Newton has a third law, which provides a clue.

Newton's Third Law of Motion: *When one body exerts a force on a second body, the second body simultaneously exerts a force equal in magnitude and opposite in direction to that of the first body.*

Newton himself used these laws to deduce an extraordinary string of observations about the solar system. As he wrote, "I now demonstrate the system of the World." To apply his ideas to the trajectory of the planets, he began by reducing each planet to a point located at the center of mass and assumed that all the planet's mass was concentrated at this point. Then, by applying his laws of motion and his new mathematics, he successfully deduced Kepler's laws of planetary motion.

He was now able to calculate the relative masses of the large planets, the Earth and the sun, and to explain a number of the curious irregularities in the motion of the moon due to the pull of the sun. He deduced that the Earth is not a perfect sphere and suggested it must be squashed between the poles due to its rotation, causing a centrifugal force. The French thought the opposite was true: that the Earth should be pointy in the direction of the poles. An expedition set out in 1733 that proved Newton—and the power of mathematics—correct.

It was an extraordinary feat. Newton's three laws meant that all motion of particles in the universe could potentially be deduced. He had come up with the seeds of a Theory of Everything. It took other scientists to grow these seeds and apply them to more complex settings. In their original form, Newton's laws were not suited to describing the motion of less rigid bodies or bodies that deform. It was the great eighteenth-century Swiss mathematician Leonhard Euler who would provide equations that generalized Newton's laws. Euler's equations could be applied more generally to something like a vibrating string or a swinging pendulum.

After Newton, more equations were developed that explained various natural phenomena. Euler produced equations for nonviscous fluids. At the beginning of the nineteenth century, French mathematician Joseph Fourier found equations to describe heat flow. Compatriots Pierre-Simon Laplace and Siméon-Denis Poisson took Newton's equations to produce more generalized equations for gravitation, which were then seen to control other phenomena like hydrodynamics and electrostatics. The behaviors of viscous fluids were described by the Navier–Stokes equations, and electromagnetism by James Clerk Maxwell's equations.

With the discovery of calculus and the laws of motion, it seemed that Newton had turned the universe into a deterministic clockwork mechanism controlled by mathematical equations. Scientists believed they had indeed discovered the Theory of Everything. In his *Philosophical Essay on Probabilities* published in 1812, the mathematician Pierre-Simon Laplace summed up most scientists' belief in the extraordinary power of mathematics to explain everything about the physical universe: "We may regard the present state of the universe as the effect of its past and the cause of its future. An intellect which at a certain moment would know all forces that set nature in motion, and all positions of all items of which nature is composed; if this intellect were also vast enough to submit these data to analysis, it would embrace in a single formula the movements of the greatest bodies of the universe and those of the tiniest atom; for such an intellect nothing would be uncertain and the future just like the past would be present before its eyes."

This view that the universe was knowable, both past, present and future, became dominant among scientists in the centuries following Newton's great opus. It seemed as if any thought of God acting in the world had been completely removed. God might be responsible for getting things up and running, but from that point on, mathematics and physics took over.

So what of my lowly die? Surely with the laws of motion at hand I can simply combine my knowledge of its geometry with information on its motion and subsequent interactions with the table to predict the outcome. I've written out the equations on my notepad, and they look pretty daunting.

Newton, as it turns out, also contemplated the problem of how to predict the role of dice. His interest was prompted by a letter he received from Samuel Pepys, who wanted Newton's advice on which option he should back in a wager he was about to make with a friend:

(1) Throwing six dice and getting at least one 6.

(2) Throwing twelve dice and getting at least two 6s.

(3) Throwing eighteen dice and getting at least three 6s.

Pepys was about to stake £10, the equivalent of £1,000 in today's money, and he was quite keen to get some good advice. Pepys's intuition was that (3) was the more likely option, but Newton replied that his calculations suggested the opposite was true. He should put his money on the first option. It wasn't his laws of motion or calculus to which Newton resorted to solve this problem but ideas developed by Fermat and Pascal.

THE FATE OF THE SOLAR SYSTEM

If Newton is my hero, then French mathematician Henri Poincaré should be the villain in my drive to predict the future. And yet I can hardly blame him for uncovering one of the most devastating blows for anyone wanting to know what's going to happen next. He was hardly very thrilled himself with the discovery, given that it cost him rather a lot of money.

Born a hundred years after Laplace, Poincaré believed, like his compatriot, in a clockwork universe, one governed by mathematical laws and utterly predictable: "If we know exactly the laws of nature and the situation of the universe at the initial moment, we can predict exactly the situation of the same universe at a succeeding moment." Understanding the world was Poincaré's prime motivation for doing mathematics. "The mathematical facts worthy of being studied are those which, by their analogy with other facts, are capable of leading us to the knowledge of a physical law," he wrote.

Newton's laws of motion had spawned an array of mathematical equations to describe the physical world, but most of them were extremely complicated. Take the equations for a gas. Think of a gas as made up of molecules crashing around like tiny billiard balls; theoretically, the future behavior of the gas should be bound up in Newton's laws of motion. But the sheer number of balls meant that an exact solution was well beyond reach. Statistical or probabilistic methods were still by far the best tool to understand the behavior of billions of molecules.

There was one situation where the number of billiard balls was reasonably small and a solution seemed tractable: the solar system. Poincaré became obsessed with predicting what lay in store for our planets as they danced their way into the future.

Because the gravitational pull of one planet on another planet at some distance from the first is the same as if all its mass were concentrated at its center of gravity, to determine the ultimate fate of the solar system one can consider planets as if they were just points in space, as Newton had done. This meant that the evolution of the solar system could be described by three coordinates for each planet, locating the center of mass in space together with three additional numbers recording the speed in each of three dimensions. The forces acting on each planet are determined by the gravitational forces exerted by each of the other planets. One just needs to apply Newton's second law to map out the course of the planets into the distant future. The trouble was that the math was extremely tricky.

Newton had solved the behavior of two planets (or a planet and a sun). They would follow elliptical paths, with their common focal point being the common center of gravity. This would repeat itself periodically to the end of time. But Newton was stumped when he introduced a third planet. Trying to calculate the behavior of a solar system consisting of the sun, the Earth, and the moon seemed simple enough, but already you are facing an equation with eighteen variables: nine for the positions and nine for the speeds of the three planets. Newton conceded that "to consider simultaneously all these causes of motion and to define these motions by exact laws admitting of easy calculation exceeds, if I am not mistaken, the force of any human mind."

King Oscar II of Norway and Sweden decided to mark his sixtieth birthday in 1889 by offering a prize for solving a problem in mathematics. There are not many monarchs around the world who would choose math problems to celebrate their birthdays, but Oscar had enjoyed the subject ever since he had excelled at it as a student at Uppsala University.

His majesty Oscar II, wishing to give a fresh proof of his interest in the advancement of mathematical science, has resolved to award a prize on January 21, 1889, to an important discovery in the field of higher mathematical analysis. The prize will consist of a gold medal of the eighteenth size bearing his majesty's image and having a value of a thousand francs, together with the sum of two thousand five hundred crowns.

Three eminent mathematicians convened to choose a number of suitable challenges and to judge the entries. One of the questions they posed was to establish mathematically whether the solar system was stable. Would it continue to turn like clockwork, or, at some point in the future, might the Earth spiral off into space and disappear from our solar system?

To answer this question it would be necessary to solve the equation that had stumped Newton. Poincaré believed that he had the skills to win the prize. One of the common tricks used by mathematicians is to attempt a simplified version of the problem first, to see if it is tractable. So Poincaré started with three bodies. This was still far too difficult, so he decided to simplify the problem further. Instead of the sun, Earth, and moon, why not try to understand two planets and a speck of dust? The two planets won't be affected by the dust particle, so he could assume, thanks to Newton's solution, that they just repeated ellipses around each other. The speck of dust, on the other hand, would experience the gravitational force of the two planets. Poincaré set about trying to describe the path traced by the speck of dust. Some understanding of its trajectory would form an interesting contribution to the problem.

Although he couldn't crack the problem completely, the paper he submitted was more than good enough to secure King Oscar's prize. He'd managed to prove the existence of an interesting class of paths that would repeat themselves, so-called periodic paths. Periodic orbits were by their nature stable because they would repeat themselves over and over, like the ellipses that two planets would be guaranteed to execute.

The French authorities were very excited that the award had gone to one of their own. The nineteenth century had seen Germany steal a march on French mathematics, so the French academicians excitedly heralded Poincaré's win as proof of a resurgence of French mathematics. Gaston Darboux, the permanent secretary of the French Academy of Sciences, declared, "From that moment on the name of Henri Poincaré became known to the public, who then became accustomed to regarding our colleague no longer as a mathematician of particular promise but as a great scholar of whom France has the right to be proud."

Preparations began for the publication of Poincaré's solution in a special edition of the Royal Swedish Academy of Science's journal *Acta Mathematica*. Then came the moment every mathematician dreads. Poincaré thought his work was safe. He'd checked every step in the proof. But just before publication, one of the editors of the journal raised a question over one of the steps in his mathematical argument.

Poincaré had assumed that a small change in the positions of the planets, a little rounding up or down here or there, was acceptable, as it would result in only a small change in their predicted orbits. It seemed a fair assumption. But there was no justification given for why this would be so. And in a mathematical proof, every step, every assumption, must be backed up by rigorous mathematical logic.

The editor wrote to Poincaré for some clarification on this gap in the proof. But as Poincaré tried to justify this step, he realized he'd made a serious mistake. He wrote to Gösta Mittag-Leffler, the head of the prize committee, hoping to limit the damage to his reputation:

> The consequences of this error are more serious than I first thought. I will not conceal from you the distress this discovery has caused me. . . . I do not know if you will still think that the results which remain deserve the great reward you have given them. (In any case, I can do no more than to confess my confusion to a friend as loyal as you.) I will write to you at length when I can see things more clearly.

Mittag-Leffler decided he needed to inform the other judges. He did so in a letter:

> Poincaré's memoir is of such a rare depth and power of invention, it will certainly open up a new scientific era from the point of view of analysis and its consequences for astronomy. But greatly extended explanations will be necessary and at the moment I am asking the distinguished author to enlighten me on several important points.

As Poincaré struggled away he soon saw that he was simply mistaken. Even a small change in the initial conditions could result in wildly different orbits. He couldn't make the approximation that he'd proposed. His assumption was wrong.

Poincaré telegraphed Mittag-Leffler to break the bad news and tried to stop the paper from being printed. Embarrassed, he wrote:

It may happen that small differences in the initial conditions produce very great ones in the final phenomena. A small error in the former will produce an enormous error in the latter. Prediction becomes impossible.

Mittag-Leffler was "extremely perplexed" to hear the news:

It is not that I doubt that your memoir will be in any case regarded as a work of genius by the majority of geometers and that it will be the departure point for all future efforts in celestial mechanics. Don't therefore think that I regret the prize. . . . But here is the worst of it. Your letter arrived too late and the memoir has already been distributed.

Mittag-Leffler's reputation was on the line for not having picked up the error before they'd publicly awarded Poincaré the prize. This was not the way to celebrate his monarch's birthday! "Please don't say a word of this lamentable story to anyone. I'll give you all the details tomorrow."

The next few weeks were spent trying to retrieve the printed copies without raising suspicion. Mittag-Leffler suggested that Poincaré should pay for the printing of the original version. Poincaré, who was mortified, agreed, even though the bill came to over 3,500 crowns, 1,000 crowns more than the prize he'd originally won.

In an attempt to rectify the situation, Poincaré set about trying to sort out his mistake, to understand where and why he had gone wrong. In 1890 Poincaré wrote a second, extended paper explaining his belief that very small changes could cause an apparently stable system suddenly to fly apart.

What Poincaré discovered, thanks to his error, led to one of the most important mathematical concepts of the twentieth century: chaos theory. It was a discovery that placed huge limits on what we humans could ever hope to know. I may have written down all the equations for my die, but what if it behaves like the planets in the solar system? According to Poincaré's discovery, if I make just one small error in recording its starting location, that error could expand into a large difference in the outcome by the time the die comes to rest on the table. So is the future of my Vegas dice shrouded behind the mathematics of chaos?

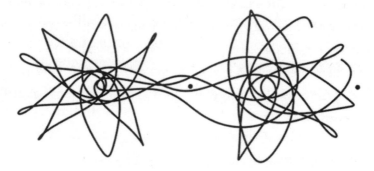

The chaotic path mapped out
by a single planet orbiting two suns

2

If nature were not beautiful it would not be worth knowing, and if nature were not worth knowing, life would not be worth living.

—Henri Poincaré

I WASTED A LOT OF TIME at university playing billiards in our student common room. I could have pretended that it was all part of my research into angles and stuff, but the truth is that I was procrastinating. It was a good way of putting off having to cope with not being able to answer that week's set of problems. But, in fact, the billiard table hides a lot of interesting mathematics in its contours, mathematics highly relevant to my desire to understand my dice.

If I shoot a ball around a billiard table and mark its path, then follow that by shooting another ball off in very nearly the same direction, the second ball will trace out a path very similar to the first ball. Poincaré had conjectured that the same principle applied to the solar system. Fire a planet off in a slightly different direction and the solar system will evolve in a very similar pattern. This is most people's intuition—if I make a small change in the initial conditions of the planet's trajectory it won't alter the course of the planet much. But the solar system seems to be playing a slightly more interesting game of billiards than the ones I played as a student.

Rather surprisingly, if you change the shape of the billiard table this intuition turns out to be wrong. If you shoot balls around a billiard table shaped like a stadium with semicircular ends but straight sides, the paths can diverge dramatically even if the balls started off heading in almost exactly the same direction. This is the signature of chaos theory: sensitivity to very small changes in the initial conditions.

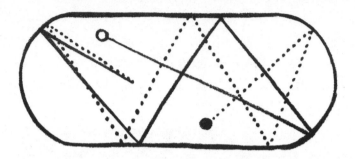

Two quickly diverging paths taken by a billiard ball around a stadium-shaped billiard table

So the challenge is to determine whether the fall of my dice can be predictable, like a conventional game of billiards, or whether we are playing a giant game of chaotic billiards.

Poincaré is generally credited as the father of chaos theory, but the sensitivity of dynamic systems to small changes was not very well known for decades into the twentieth century. It really took the redis-covery of the phenomenon by scientist Edward Lorenz, who like Poin-caré thought he'd made some mistake, before the ideas of chaos theory became more widely known.

While working as a meteorologist at the Massachusetts Institute of Technology in 1963, Lorenz had been running equations for the change of temperature in a dynamic fluid on his computer when he decided he needed to rerun one of his models for longer. So he took some of the data that had been output earlier in the run and re-entered it, expecting to be able to restart the model from that point.

When he returned from coffee, he discovered to his dismay that the computer hadn't reproduced the previous data and that it had very quickly generated a wildly divergent prediction. At first he couldn't understand what was happening. If you input exactly the same numbers into an equation, you don't expect to get a different answer at the other end. It took him a while to realize what was going on: he hadn't input the same numbers. The computer printout of the data he'd used had only printed the numbers to three decimal places, while it had been calculating using the numbers to six decimal places.

Even though the numbers were different, they differed only in the fourth decimal place. You wouldn't expect this to make that big a difference, but Lorenz was struck by the impact of such a small difference on the resulting data. Here are two graphs created using the same equation, where the data that are put into the equations differ very slightly. One graph uses the input data 0.506127 and the second graph approximates this to 0.506. Although the graphs start out following similar paths, they very quickly behave completely differently.

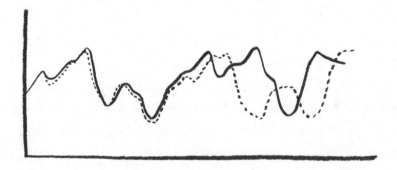

The model that Lorenz was running was a simplification of weather models that analyzed how the flow of air behaves when subjected to differences in temperature. His rediscovery of how small changes in starting conditions can have such a big impact on future outcomes would have huge implications for our attempts to use mathematical equations to predict the future. As Lorenz wrote, "Two states that were imperceptibly different could evolve into two considerably different states. Any error in the observation of the present state—and in a real system, this appears to be inevitable—may render an acceptable prediction of the state in the distant future impossible."

When Lorenz sought to explain his findings to a colleague, he was told, "Edward, if your theory is correct, one flap of a seagull's wings could alter the course of history forever." The seagull would eventually be replaced by the now famous butterfly when Lorenz presented his findings in 1972 at the American Association for the Advancement of Science in a paper titled "Does the Flap of a Butterfly's Wings in Brazil Set off a Tornado in Texas?"

Curiously, both the seagull and the butterfly might have been preempted by the grasshopper. It seems that already in 1898 Professor W. S. Franklin had realized the devastating effect that the insect community could have on the weather. In a book review, he posited, "An infinitesimal cause may produce a finite effect. Long-range detailed weather prediction is therefore impossible, and the only detailed prediction which is possible is the inference of the ultimate trend and character of a storm from observations of its early stages; and the accuracy of this prediction is subject to the condition that the flight of a grasshopper in Montana may turn a storm aside from Philadelphia to New York!"

This is an extraordinary position to be in. Science offers a completely deterministic description of the evolution of many dynamic systems like the weather. And yet in many cases we are denied access to its predictions, as any measurement of the location or wind speed of a particle is inevitably going to be an approximation of its true conditions.

The National Weather Service, when making weather predictions, takes the data recorded by weather stations dotted across a region and then, instead of running equations on these data, the meteorologists do several thousand runs, varying the data over a range of values. The predictions stay close for a while, but by about five days into the future the results have often diverged so wildly that one set of data predicts a heat wave while a few changes in the decimal places of the data result in drenching rain.

"There is a maxim which is often quoted, that 'The same causes will always produce the same effects,'" wrote the great Scottish scientist James Clerk Maxwell in his book *Matter and Motion*, published in 1877. "There is another maxim which must not be confounded with this, which asserts that 'Like causes produce like effects.' This is only true when small variations in the initial circumstances produce only small variations in the final state of the system." The discovery of chaos theory in the twentieth century revealed this maxim to be false.

Of course, there are times when small changes don't alter the course of the equations dramatically, like the paths in the classic billiard table. So how can you know the point when you can't know what is going to happen next? Robert May discovered a beautiful example when he analyzed the equations for population growth.

Born in Australia in 1938, May had originally trained as a physicist working on superconductivity, but his academic work took a dramatic turn when he was exposed in the late 1960s to the newly formed movement for social responsibility in science. His attention shifted from the behavior of collections of electrons to the more pressing questions of the behavior of population dynamics in animals. Biology, at the time, was not a natural environment for the mathematically minded, but following May's work that would change. This fusion of the hardcore mathematical training he'd received as a physicist with a new interest in biology led to his great breakthrough.

In a paper in *Nature* called "Simple Mathematical Models with Very Complicated Dynamics," published in 1976, May explored the dynamics of a mathematical equation describing population growth from one season to the next. He revealed how even a quite basic equation can produce extraordinarily complex results. His equation for population dynamics wasn't some complicated differential equation but a simple, discrete feedback equation that anyone with a calculator could explore.

May's Feedback Equation for Population Dynamics

Consider a population of animals whose numbers can vary between 0 and some hypothetical maximum value that I will call N. Given some fraction Y (lying between 0 and 1) of that maximum, the equation determines the proportion of the population that will survive in the next season after reproduction and competition for food. Let's suppose that each season the reproduction rate is given by a number r. If the fraction of the maximum population that survived to the end of the season was Y, the next generation would swell to $r \times Y \times N$.

But not all of these new animals will survive. The equation determines that the fraction that will not survive is also given by Y. So out of the $r \times Y \times N$ animals that start the season, $Y \times (r \times Y \times N)$ die. So the total left at the end of the season is $(r \times Y \times N) - (r \times Y^2 \times N) = [r \times Y \times (1-Y)] \times N$, which means that the fraction of the maximum population that exists in the current season is $r \times Y \times (1-Y)$.

Essentially, the model assumes that at the end of each season the surviving population is multiplied by a constant factor, called r (the reproduction rate) to produce the number of animals that will be alive at the beginning of the next season. But there aren't enough resources for all of them to survive. The equation then calculates how many of these animals will make it until the end of the season. The resulting number is then multiplied by r again for the next generation. The fascinating property of this equation is that its behavior really depends only on the choice of r, the reproduction rate. Some choices of r lead to extremely predictable behaviors. I can know exactly how the numbers will evolve. But there is a threshold beyond which I lose control. Knowledge is no longer within reach; the addition of one extra animal into the mix can result in dramatically different population dynamics.

May discovered that if r lies between 1 and 3, the population eventually stabilizes. Regardless of the initial conditions, the numbers will gradually tend to be a fixed value. It's like playing billiards on a table with a sinkhole in the middle. However I shoot the ball, it eventually finds its way to the bottom of the sinkhole.

For r above 3, we still see a region of predictable behavior, but of a slightly different character. If r remains between 3 and $1 + \sqrt{6}$ (approximately 3.44949) the population dynamics ping-pong between two values that depend on r. As r passes $1 + \sqrt{6}$, we see the population dynamics changing character again. For r between $1 + \sqrt{6}$ and 3.54409 (or, more precisely, the solution of a polynomial equation of degree 12), there are four values that the population periodically cycles through. As r gets bigger, we get eight values, then sixteen, and so on. As r climbs, the number of different values doubles each time until we hit a threshold moment, when the dynamic flips from periodic to chaotic.

When May first explored this equation, he admitted that he frankly hadn't a clue what was going on beyond this point—he had a blackboard outside his office in Sydney on which he offered a prize of ten Australian dollars to anyone who could explain the behavior of his equation. He wrote on the blackboard, "It looks like a mess."

He got his answer on a visit to Maryland. This was, in fact, when the term "chaos" was actually coined. In a seminar, he explained his findings but admitted he'd hit a point beyond which he didn't know what the hell was happening. In the audience was a mathematician who did know. Jim Yorke had never seen the doubling behavior, but he knew exactly what was going on in this higher region. And it was what he called chaos.

Beyond $r = 3.56995$ (or, more precisely, the limit point of a system of equations of increasing degree), the result becomes very sensitive to what the initial population looks like. Change the initial number of animals by a minute amount and a totally different result can ensue.

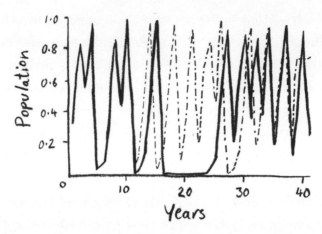

Two populations with *r* = 4 that start off with a difference of just one animal in a thousand. Although they start behaving similarly, by year 15 they are demonstrating very different behaviors.

As I turn up the dial on *r*, there can still be pockets of regular behavior, as Jim Yorke had discovered. For example, take *r* = 3.627 and the population becomes periodic again, bouncing around between six different values. Keep dialing *r* up and the six changes to twelve, which becomes twenty-four, doubling each time until chaos strikes again.

Bob May recognized just what a warning shot such a simple system was to anyone who thought they knew it all: "Not only in research, but in the everyday world of politics and economics, we would be better off if more people realized that simple systems do not necessarily possess simple dynamic properties."

THE POLITICS OF CHAOS

Bob May is currently practicing what he preaches. Or perhaps I should say Lord May of Oxford, as I was corrected by a man in a top hat who greeted me at the door to the entrance of the House of Lords. May has in recent years combined his scientific endeavors with energetic political activism. He now sits as a cross-party member of the House of Lords, where I popped in for lunch to find out how he was faring in his mission to alert politicians to the impact of chaotic systems on society.

Ushered through the entrance to the Lords by the man in the top hat and policemen with machine guns, I found May waiting for me on the other side of metal detectors and X-ray machines. May has no truck with all these formal titles and, in his earthy Australian manner, still insists on being called Bob. "I'm afraid I messed up and already ate lunch but I'll come and eat cake while you get some lunch," he said with a guileless smile. As I ate my fish he consumed an enormous piece of chocolate cake. At seventy-nine, May is as energetic and engaged as ever, and he was rushing off after his second lunch to a select committee discussing the impact of a new rail link between London and northwest England. Before joining the Lords, May was chief scientific adviser both to John Major's Conservative government and Tony Blair's Labor government. I asked how tricky a balancing act such a political position is for a man who generally is not scared to tell it like it is.

"At the interview I was told that there would be occasions where I would be called upon to defend the decisions of a minister and how would I feel about that? I said that I would never under any circumstances deny a fact. On the other hand, I'm fairly good at the kind of debating competition where you're given a topic and according to a flip of a coin you've got to argue for either side of the debate. So I said I'd be happy explaining why the minister's choice was arrived at. I simply wouldn't agree to endorse it if it wasn't right."

A typical mathematician's response. Set up the minister's axioms and then demonstrate the proof that led to the conclusion—a judgment-free approach. That's not to say that May isn't opinionated and prepared to give his own views on the subject at hand.

I was curious as to how governments deal with the problems that chaos theory creates when trying to make policy decisions. How do politicians cope with the challenges of predicting or manipulating the future, given that we can have only partial knowledge of the systems being analyzed?

"I think that's rather a flattering account of what goes on here," he said. "With some notable exceptions it's mostly a bunch of very egotistical people, very ambitious people, who are primarily interested in their own careers."

What about May personally? What impact did the discoveries he'd made have on his view of science's role in society?

"It was weird. It was the end of the Newtonian dream. When I was a graduate student it was thought that with better and better computer power we would get better and better weather predictions because we knew the equations and we could make more realistic models of the Earth." But May is cautious not to let the climate change deniers use chaos theory as a way to undermine the debate.

"Not believing in climate change because you can't trust weather reports is a bit like saying that because you can't tell when the next wave is going to break on Bondi beach you don't believe in tides."

May likes to quote a passage from Tom Stoppard's play *Arcadia* to illustrate the strange tension that exists between the power of science to know some things with extraordinary accuracy and chaos theory, which denies us knowledge of many parts of the natural world. One of the protagonists, Valentine, declares, "We're better at predicting events at the edge of the galaxy or inside the nucleus of an atom than whether it'll rain on auntie's garden party three Sundays from now." May jokes that his most cited works are not the high-profile academic papers he's published in prestigious scientific journals like *Nature*, but the program notes he wrote for Stoppard's play when it was first staged at the National Theater in London. "It makes a mockery of all these citation indexes as a way of measuring the impact of scientific research."

So what are the big open questions of science that May would like to know the answer to? Consciousness? An infinite universe?

"I think I'd look at it in a less grand way, so I'd look at it more in terms of the things I am working on at the moment. Largely by accident I've been drawn into questions about banking."

That was a surprise. The question of how to create a stable banking system seemed very parochial, but May has recently been applying his models of the spread of infectious diseases and the dynamics of ecological food webs to understanding the banking crisis of 2008. Working with Andrew Haldane at the Bank of England, he has been considering the financial network as if it were an ecosystem. Their research has revealed how financial instruments intended to optimize returns with seemingly minimal risk can cause instability in the system as a whole.

May believes that the problem isn't necessarily the mechanics of the market itself. It's the way small things in the market are amplified and perverted by how humans interact with them. For him, the most worrying thing about the banking mess is getting a better handle on this contagious spreading of worry.

"The challenge is: How do you put human behavior into the model? I don't think human psychology is mathematizable. Here we are throwing dice with our future. But if you're trying to predict the throw of the dice, then you want to know the circumstance of who owns the dice."

That was something I hadn't taken into account. Perhaps I should factor in who sold me the casino die in the first place.

"I think many of the major problems facing society are outside the realm of science and mathematics," he said. "It's the behavioral sciences that are the ones we are going to have to depend on to save us."

Looking around the canteen at the House of Lords, you could see the sheer range and complexity of human behavior at work. It makes the challenge of mathematizing even the interactions in this tiny microcosm of the human population well nigh impossible. As the French historian Fernand Braudel explained in a lecture on history he gave to his fellow inmates in a German prison camp near Lübeck during World War II, "An incredible number of dice, always rolling, dominate and determine each individual existence." Although each individual throw is unpredictable, patterns still emerge in the long-range behavior of many throws of the dice. In Braudel's view, this is what makes the study of history possible: "History is indeed 'a poor little conjectural science' when it selects individuals as its objects . . . but much more rational in its procedure and results when it examines groups and repetitions."

May believes that understanding the history and origins of the collection of dice that make up the human race is not as straightforward as Braudel makes out. To his mind, it's not at all clear that we can unpick how we got to this point in our evolutionary journey.

"I'll tell you one of the questions that I think is a particularly interesting one: trying to understand our evolutionary trajectory as humans on our planet. Is the trajectory we seem to be on what happens on all or most planets, or is it the result of earlier fluctuations in the chaos which took us on this trajectory rather than another? Will we ever know enough to be able to ask whether the disaster we seem to be heading for is inevitable or whether there are lots of other planets where people are more like Mr. Spock, less emotional, less colorful, but much more detached and analytical?"

Until we discover other inhabited planets and can study their trajectories, it's difficult to assess whether evolution inevitably leads to mismanaged ecosystems based on just one dataset called Earth.

"The question of whether where we're heading is something that happens to all inhabited planets or whether there are other planets where it doesn't happen is something I think we'll never know." And with that, May polished off the last few crumbs of his chocolate cake and plunged back into the chaos of the select committees and petty politics of Westminster.

May's last point relates to the challenge that chaos theory poses for knowing something about the past as much as the future. At least with the future we can wait and see what the outcome of chaotic equations produces. But trying to work backward and understand what state our planet was in to produce the present is equally if not more challenging. The past, even more than the future, is probably something we can never truly know.

LIFE: A CHANCE THROW OF THE DICE?

May's pioneering research explored the dynamics of a population as it went from season to season. But what determines which animals survive and which will die before reproducing? According to Darwin, this is simply down to a lucky roll of the evolutionary dice.

Darwin's model for the evolution of life on Earth is based on the idea that once you have organisms with DNA, then the offspring of these organisms share the DNA of their parent organisms. Parts of the genetic code in the DNA can undergo random mutations. These are essentially down to the chance throw of the evolutionary dice. But there is a second important strand to Darwin's proposal, which is the idea of natural selection. Some of those random changes will give offspring an increased chance of survival, while others will result in a disadvantage. The point of evolution by natural selection is that the organism with the advantageous change will be more likely to survive long enough to reproduce.

Suppose, for example, that I start with a population of giraffes with short necks. The environment of the giraffes changes such that there is more food in the trees, so that any giraffe born with a longer neck is going to have a better chance of survival. Let's suppose that I throw my Vegas die to determine the chance of a mutation for each giraffe born in the next generation following this environmental change. A roll of a 1, 2, 3, 4, or 5 condemns the giraffe to a neck of the same size or shorter, while a throw of a 6 corresponds to a chance mutation that causes a longer neck. The lucky longer-necked giraffes get the food, and the shorter-necked giraffes don't survive to repro-duce. So it is just the longer-necked giraffes that get the chance to pass on their DNA.

In the next generation the same thing happens. Roll a 1, 2, 3, 4, or 5 on the die and the giraffe doesn't grow any taller than its parents. But another 6 and the giraffe's neck grows a bit more. The taller gi-raffes survive again. The environment favors the giraffes that have thrown a 6. Each generation ends up a bit taller than the last genera-tion until there comes a point when it is no longer an advantage to grow any further.

The combination of chance and natural selection results in our seeing more giraffes with ancestors that all threw 6s. In retrospect it looks like amazing serendipity that you would see so many 6s in a row. But the point is that you don't see any of the other rolls of the dice because they don't survive. What looks like a rigged game is just the result of the combination of chance and natural selection. There is no grand design at work. The run of consecutive 6s isn't a lucky streak; it is the only thing we would expect to see from such a model.

It's a beautifully simple model, but, given the complexity of the changes in the environment and the range of mutations that can occur, this simple model can produce extraordinary complexity, which is borne out by the sheer variety of species that exist on Earth. Although the model is simple, it is frustratingly inadequate at making predic-tions. One of the reasons I never really fell in love with biology is that there seemed to be no way to explain why we got cats and zebras out of this evolutionary model and not some other strange selection of ani-mals. It all seemed so arbitrary, so random. But is that really right?

There is an interesting debate going on in evolutionary biology about how much chance there is in the outcomes we are seeing. If we rewound the story of life on Earth to some point in the past and threw the dice again, would we see very similar animals appearing or could we get something completely different? This is the question that May raised at the end of our lunch.

It does appear that some parts of our evolutionary process seem inevitable. It is striking that throughout evolutionary history, the eye evolved independently fifty to a hundred times. This is strong evidence for the fact that the different rolls of the dice that have occurred across different species seem to have produced species with eyes regardless of what is going on around them. Lots of other examples illustrate how some features, if they are advantageous, seem to rise to the top of the evolutionary swamp. This is illustrated every time you see the same feature appearing more than once in different parts of the animal kingdom. Dolphins and bats, for example, use echolocation, but they evolved this trait independently at very different points on the evolutionary tree.

But it isn't clear how far these outcomes are guaranteed by the model. If there is life on another planet, will it look anything like the life that has evolved here on Earth? This is one of the big open questions in evolutionary biology. As difficult as it may be to answer, I don't believe it qualifies as something we can never know. It may remain something we will never know, but there is nothing by its nature that makes it unanswerable.

Are there other great unsolved questions of evolutionary biology that might be contenders for things we can never know? For example, why, 542 million years ago, at the beginning of the Cambrian period, was there an explosion in the diversity of life on Earth? Before this moment life consisted of single cells that collected into colonies. But over the next twenty-five million years, a relatively short period on the scale of evolution, there is a rapid diversification of multicellular life that ends up resembling the diversity that we see today. An explanation for this exceptionally fast pace of evolution is still missing. This is in part due to lack of data from that period. Can we ever recover that information, or could this always remain a mystery?

Chaos theory is usually a limiting factor in what we can know about the future. But it can also imply limits on what we can know about the past. We see the results, but deducing the cause means running the equations backward. Without complete data, the same principle applies backward as forward. We might find ourselves at two very divergent starting points that can explain very similar outcomes. But we'll never know which of those origins was ours.

One of the big mysteries in evolutionary biology is how life got going in the first place. The game of life may favor runs of 6s on the roll of the evolutionary dice, but how did the game itself evolve? Various estimates have been proposed for the chances of everything lining up to produce molecules that replicate themselves. In some models, the origin of life is equivalent to nature having thrown thirty-six dice and getting them all to land on 6. For some, this is proof of the existence of God or of some form of a grand designer to rig the game. But this is to misunderstand the huge time scale that we are working on.

Miracles do happen . . . given enough time. Indeed, it would be more striking if we didn't get these anomalies. The point is that the anomalies stick out. They get noticed, while the less exciting rolls of the dice are ignored.

The lottery is a perfect test bed for the occurrence of miracles in a random process. On September 6, 2009, the following six numbers were the winning numbers in the Bulgarian state lottery: 4, 15, 23, 24, 35, 42. Four days later, the same six numbers came up again. Incredible, you might think. The government in Bulgaria certainly thought so and ordered an immediate investigation into the possibility of corruption. But what it failed to take into account is that each week, across the planet, different lotteries are being run. They have been running for decades. If you do the math, it would be more surprising not to see such a seemingly anomalous result.

The same principle applies to the conditions for producing self-replicating molecules in the primeval soup that made up the Earth before life emerged. Mix together plenty of hydrogen, water, carbon dioxide, and some other organic gases and subject them to lightning strikes and electromagnetic radiation, and already experiments in the lab show the emergence of organic material found only in living things. No one has managed to spontaneously generate anything as extraordinary as DNA in the lab. The chances of that are very small.

But that's the point. Given the billion billion or so possible planets available in the universe on which to try out this experiment, together with a billion or so years to let the experiment run, it would be more striking if that outside chance of creating something like DNA didn't happen. Keep rolling thirty-six dice on a billion billion different planets for a billion years and you'd probably get one roll with all thirty-six dice showing 6. Once you have a self-replicating molecule, it has the means to propagate itself, so you only need to get lucky once to kick off evolution.

THE FRACTAL TREE OF LIFE

Our problem as humans is that we have not evolved minds able to navigate very large numbers. Probability is something we have little intuition for. But it's not only the mathematics of probability that is at work in evolution. The evolutionary tree itself has an interesting quality that it shares with chaos theory. This quality is known as a fractal.

The fractal evolutionary tree

The evolutionary tree is a graphic representation of the evolution of life on Earth. Each time the tree branches, this represents the evolution of a new species. If a branch terminates, this means the extinction of that species. The nature of the tree is such that the overall shape seems to be repeated on smaller and smaller scales. This is the characteristic feature of a shape mathematicians call a fractal. If you zoom in on a small part of the tree it looks remarkably like the large-scale structure of the tree. This self-similarity means that it is very difficult to tell at what scale we are looking at the tree. This is the classic characteristic of a fractal.

Fractals are the geometric signature of a chaotic system, so it is suggestive of chaotic dynamics at work in evolution: small changes in the genetic code can result in huge changes in outcome. This model isn't necessarily a challenge to the idea of convergence, as there can still be points in chaotic systems toward which the model tends to evolve. Such points are called attractors. But it certainly questions whether if you reran evolution it would look anything like what we've got on Earth today. The evolutionary biologist Stephen Jay Gould contended that if you were to rerun the tape of life, you would get very different results. This is what you would expect from a chaotic system. Just as with the weather, very small changes in the initial conditions can result in dramatically different outcomes.

Gould introduced the idea of punctuated equilibria to capture the fact that species seem to remain stable for long periods and then undergo what appears to be quite rapid evolutionary change. This has been shown to be a feature of chaotic systems. The implications of chaos at work in evolution could well fall under the umbrella of things we cannot know.

Will we ever know whether humans were destined to evolve? An analysis of DNA in different animals has given us exceptional insights into the way animals have evolved in the past. The fossil record, although incomplete in places, has also given us a way to know our origins. But given the time scales involved in evolution, it is impossible to experiment and rerun the tape of life and see whether something different could have happened. As soon as we find life on other planets (*if we do*), this will give us new sample sets to analyze. But until then all is not lost. Just as the National Weather Service doesn't have to run real weather to make predictions, computer models can illustrate different possible outcomes of the mechanism of evolution by speeding up time. But the model will only be as good as our hypotheses. If we've got the model wrong, it won't tell us what is really happening in nature.

Computer models such as these hold the key to answering the question Poincaré first tackled when he discovered chaos: Will there even be a stable Earth orbiting the sun for evolution to continue playing its game of dice? How safe is our planet from the vagaries of chaos? Is our solar system stable and periodic, or do I have to worry about a grasshopper disrupting our orbit around the sun?

A BUTTERFLY CALLED MERCURY

Poincaré wasn't able to answer the King of Sweden's question about the solar system, that is, whether it would remain in a stable equilibrium or fly apart in a catastrophic exhibition of chaotic motion. His discovery that some dynamic systems can be sensitive to small changes in data opened up the possibility that we may never know the precise fate of the solar system much in advance of any potentially devastating scenario unfolding.

It is possible that the solar system is in a safe, predictable region of activity, but the evidence suggests we can't console ourselves with this comforting mathematical hope. Recent computer modeling has provided us with new insights, which reveal that the solar system is indeed within a region dominated by the mathematics of chaos. We can now measure how big an effect a small change will have on the outcome of a closed system using something called the Lyapunov expo-nent. If the Lyapunov exponent is positive, it means that if I make a small change in the initial conditions then the distance between the paths will diverge exponentially.

Using this new equation, several groups of scientists have con-firmed that our solar system is indeed chaotic. They have calculated that the distance between two initially close orbital solutions increases by a factor of ten every ten million years. This is certainly on a differ-ent timescale to our inability to predict the weather. Nevertheless, it means that I can have no definite way of knowing what will happen to the solar system over the next five billion years.

If you're wondering in despair whether we can know anything about the future, then take heart in the fact that mathematics isn't completely hopeless at making predictions. There is an event that the equations guarantee will occur if we make it to five billion years from now, but it's not good news. At this point, the sun will run out of fuel and evolve into a red giant that will engulf Earth and the other planets in our solar system. But until this solar blowout, I am faced with try-ing to solve chaotic equations if I want to know which planets will still be around to see that red giant.

If I want to know what will happen, I have no choice but to run simulations in which I vary the precise locations and speeds of the planets. The forecast is in some cases rather frightening. In 2009 French astronomers Jacques Laskar and Mickael Gastineau ran sev-eral thousand models of the future evolution of our solar system. Their experiments have identified a potential butterfly: Mercury.

The simulations start by feeding in the records we have of the posi-tions and velocities of the planets to date, but it is difficult to know these with one hundred percent accuracy. So each time they run the simula-tion they make small changes to the data. Because of the effects of chaos theory, a small change could result in a large deviation in the outcomes.

For example, astronomers know the dimensions of the ellipse of

Mercury's orbit to an accuracy of several meters. Laskar and Gastineau ran 2,501 simulations varying these dimensions over a range of less than a centimeter. Even this small perturbation resulted in startlingly different outcomes for our solar system.

If the solar system were to be ripped apart, you might expect one of the big planets, like Jupiter or Saturn, would be the culprit. But the orbits of the gas giants are extremely stable. It's the rocky terrestrial planets that are the troublemakers. In one percent of simulations, Laskar and Gastineau found that tiny Mercury posed the biggest risk. The models show that Mercury's orbit could start to extend due to a certain resonance with Jupiter, with the possibility that Mercury could collide with its closest neighbor, Venus. In one simulation, a close miss was enough to throw Venus out of kilter, with the result that Venus collides with Earth. Even close encounters with the other planets would be enough to cause such tidal disruption that the effect would be disastrous for life on our planet.

This isn't simply a case of abstract mathematical speculation. Evidence of such collisions has been observed in the planets orbiting the binary star Upsilon Andromedae. Their current strange orbits can be explained only by the ejection of an unlucky planet sometime in the star's past. But before we head for the hills, the simulations reveal that it will take several billion years before Mercury might start to misbehave.

INFINITE COMPLEXITY

What of my more modest goal of predicting the throw of my die? Laplace would have said that provided I can know its dimensions, the distribution of its atoms, the speed at which it is launched, and its relationship to its surrounding environment, theoretically the calculation is possible. But the discoveries of Poincaré and those who followed have revealed that just a few decimal places could be the difference between a 6 or a 2. The die is designed to have only six different outcomes, yet the input data range over a potentially continuous spectrum of values. The question is whether the dynamics of the die are truly chaotic—or could they be simpler than one might expect? If I vary the angle at which the die leaves my hand, is there a moment when the outcome flips from a 6 to a 2, or is it much more sensitive to small changes?

A Polish research team recently analyzed the throw of a die mathematically, and by combining this with the use of high-speed cameras they have revealed that my die may not be as chaotic and unpredictable as I feared. It just depends on the conditions of the table onto which you are throwing your die. The research group consists of a father-and-son team, Tomasz and Marcin Kapitaniak, together with Jaroslaw Strzalko and Juliusz Grabski, and they are based in Lódź. The model they considered, published in a paper in the journal *Chaos* in 2012, assumes that the die is perfectly balanced, like the one I brought back from Vegas. Air resistance, it turns out, can be ignored, as it has very little influence on our cube as it tumbles through the air. When it hits the table a certain proportion of its energy is dissipated, so that after a number of bounces it has lost all kinetic energy and comes to a rest. (Friction on the table is also key, as the die is likely to slide only in the first few bounces, but the model explored by the Polish team assumed a frictionless surface, as the dynamics get too complicated to handle when friction is present. So imagine throwing your dice onto an ice rink.)

I'd already written down equations based on Newton's laws of motion to calculate the dynamics of the die as it flies through the air. In the hands of the Polish team they turn out not to be too complicated, but the equations for the change in dynamics after the impact with the table are pretty frightening, taking up ten lines of their paper.

The paper includes pictures that describe the outcome of their experiments. The Polish team investigated the effect of varying two parameters: the height from which the die was launched and the angular velocity around one of the axes. They gave each face of the die its own color and constructed a picture that took each coordinate in the diagram and colored it according to the face of the die that resulted from launching it within the parameters corresponding to that coordinate. So if the resulting picture is fractal—meaning that however much I zoom in, I still see regions containing at least two colors—then the outcome of the roll of the die is impossible to predict. The Polish team tried throwing their dice on different types of tables. They discovered that if the amount of energy dissipated on impact with the table is quite high, the picture of the outcome does not have a fractal quality. This means that if one can settle the initial conditions with appropriate accuracy, the outcome can be predictable and repeatable. This predictability implies that more often than not, the dice will land on the face that was lowest when they were launched. A die that is fair when static may actually be biased when one adds in its dynamics.

As the table becomes more rigid, resulting in less energy being dissipated and hence more bouncing, I start to see a fractal quality emerging.

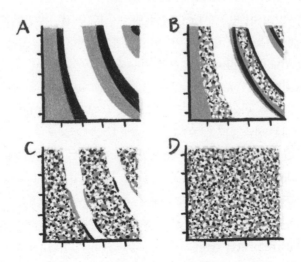

Moving from (a) to (d), the table dissipates less energy, resulting in a more fractal quality for the outcome of the dice.

So the bottom line is look for a casino whose craps tables have thick felt that might absorb the energy of the die as it hits the table. The less energy that is dissipated on impact with the table, the more chaotic its resulting behavior and the more it seems that the outcome recedes back into the hands of the gods.

There is an important message here. Dynamical systems like the weather, population growth, the economy, or the fall of my die can have regions that are very predictable and knowable. Throw the die on a table that causes the die to lose energy and the behavior of the die is not chaotic. Approximate knowledge of how the die is launched will be enough to calculate its outcome. But there are other conditions under which these dynamical systems become chaotic and unknowable. The trick is knowing for any dynamical system the moment when it flips from being predictable to chaotic, knowing when you can't know what is going to happen next.

DOES GOD PLAY DICE?

In the deep past, gods weren't considered to be supernatural intelligences living outside the system. They were rivers, the wind, fire, the sun—things that could not be predicted or controlled. They lived in the realm of chaos. Twentieth-century mathematics has revealed that these ancient gods are still with us. There are natural phenomena that will never be tamed and known. Chaos theory asserts that I cannot know the future of certain systems because they are too sensitive to small inaccuracies. Because we can never have complete knowledge of the present, chaos theory denies us access to the future.

That's not to say that all futures are unknowable. Very often we are in regions that aren't chaotic, where small fluctuations have little effect. This is why mathematics has been so powerful in helping us to predict and plan. The power of mathematical equations has allowed us to land spaceships on other planets, predict the paths of deadly typhoons on Earth, and model the effects of deadly viruses, allowing us to take action before they become a pandemic. But at other times we cannot accurately predict or control outcomes.

Some religious commentators who know their science and seek to offer a scientific explanation for how a supernatural intelligence could act in a scientific world have intriguingly tried to use the gap that chaos provides as a space for this intelligence to affect the future. One of these is the quantum physicist John Polkinghorne. Based at the University of Cambridge, Polkinghorne is a rare man who combines both the rigors of a scientific education with years of training to be a Christian priest. I will return to Polkinghorne in the Third Edge, when I explore the unknowability inherent in his own scientific field of quantum physics, but he has also been interested in the gap in knowledge that the mathematics of chaos theory provides as an opportunity for God to influence the future course of humanity.

Polkinghorne has proposed that a supernatural intelligence can still act without violating the laws of physics via the indeterminacies implicit in chaos theory. Chaos theory says that we can never know the setup precisely enough to be able to run deterministic equations, and hence there is room in Polkinghorne's view for divine intervention, to tweak things in such a way as to remain consistent with our partial knowledge but still influence outcomes.

Polkinghorne is careful to stress that to use infinitesimal data to effect change requires a complete holistic top-down intervention. This is not a God in the details but an all-knowing God. Given that chaos theory means that even the location of an electron on the other side of the universe could influence the whole system, we need to have complete, holistic knowledge of the system—the whole universe—to be able to steer things. We cannot successfully isolate a part of the universe and hope to make predictions based on that part. So it would require knowledge of the whole to act via this tiny chink.

Polkinghorne's take on how to square the circle of determinism and influence is to use the gap between epistemology and ontology—in other words, between what we know and what is true. It is important to recognize that chaos theory is deterministic. It is not random. Repeat the experiment with exactly the same starting conditions and the result will be the same each time. But since we cannot know with complete certainty the state of the universe at this moment in time, from our perspective there is no determinacy. Many different scenarios coincide with our impartial description of what we currently know about how the universe is set up. Polkinghorne's contention is that at any point in time this gives God the chance to intervene and shift the system between any of these scenarios without our being aware of the shift. As we have seen, chaos theory means that these small shifts can still result in hugely different outcomes. Polkinghorne is careful to assert that there is change only in information, not energy. The goal here is not to violate any rules of physics. As Polkinghorne says, "The succession of the seasons and the alternations of day and night will not be set aside."

Even if you think this is rather fanciful (as I do), a similar principle is key to our own feeling of agency in the world. Free will is related ultimately to the deep desire to challenge a belief in a reductionist philosophy. It makes sense to create a narrative in which we have free will because that is how we experience life. If things were obviously deterministic, we wouldn't think that we had any meaningful choice. From a human perspective, we feel like we make meaningful choices every day. But perhaps that feeling is just a result of chaos theory masking our ability to see determinism at work in the world.

It is striking that Newton, the person who led us to believe in a clockwork deterministic universe, also felt that there was room in the equations for God's intervention. He wrote of his belief that God would sometimes have to reset the universe when things looked like they were going off course. He got into a big fight with his German mathematical rival, Gottfried Leibniz, who couldn't see why God wouldn't have set things up perfectly from the outset. In Leibniz's view, "Sir Isaac Newton and his followers have also a very odd opinion concerning the work of God. According to their doctrine, God Almighty wants to wind up his watch from time to time: otherwise it would cease to move. He had not, it seems, sufficient foresight to make it a perpetual motion."

Newton's calculus gave me a feeling that I could know the future, that I could shortcut the wait for it to become the present. But the mathematics of the twentieth century revealed that theory doesn't necessarily translate into practice. Even if Laplace is correct that complete knowledge of the current state of the universe should lead to complete knowledge of the future, I will never have access to that complete knowledge. The shocking revelation of twentieth-century chaos theory is that even an infinitesimally small approximation can change everything. The divergent paths of the chaotic billiard table mean that since we can never know which path we are on, our future is not predictable.

The mathematics in which I had placed so much faith revealed its own limitations. But the situation is not entirely hopeless. Many times equations are not sensitive to small changes and hence do give us access to predictions about the future. After all, this is how we landed a spaceship on a passing comet.

It is striking to listen to Laplace on his deathbed. With only a finite amount of time to go, he admitted, "What we know is little, and what we are ignorant of is immense." It isn't just the outward behavior of planets that is unknowable. Probing deep inside my casino die reveals another challenge to Laplace's belief in a clockwork universe.

SECOND EDGE

MATTER

3

Everyone takes the limits of his own vision for the limits of the world.

—Arthur Schopenhauer

WHEN I STARTED AT THE LOCAL SCHOOL, I remember my music teacher asking the class whether anyone wanted to learn a musical instrument. Three of us put up our hands. The teacher led us into the storeroom cupboard to see what instruments were available. The cupboard was bare except for three trumpets stacked up on top of each other.

"It looks like you're learning the trumpet."

I don't regret the choice (even if there wasn't one). I had a great time playing in the local town band and larking around in the brass section of the county orchestra as we counted bars of rest. But I used to look over with a little envy at the string section, which seemed to be playing all the time, getting all the good tunes. A few years ago, during a radio interview, I was asked what new musical instrument I would choose to learn, given the opportunity, and which piece of music I would aspire to play.

"The cello. Bach's suites," I said without thinking.

The question has been nagging at the back of my mind since that interview—could I learn to play those beautiful cello suites? Perhaps it was too late to pick up such new skills, but I needed to know. So I bought a cello. It sits behind me as I write. When I need a break from analyzing equations, I massacre one of the gigues from the first suite for cello. I can feel Bach turning in his grave, but I am enjoying myself.

One of the fun things about the cello is the possibility of sliding your finger up the string to create a continuous glissando of notes. This is not something I can do on my trumpet, which is an instrument of discrete notes corresponding to different combinations of fingers. It turns out that this tension between the continuous glissando of the cello and the discrete notes of the trumpet is relevant to our attempt to predict the behavior of my die.

To predict how it will land I need to know what it is made from. Denser acetate in one corner will lead to one side of the die being favored over the others. So if I am going to attempt to apply Newton's laws to my die as it tumbles through the air, I need to know how it is put together. Is it a continuous structure, or, if I look closely, is it made up of discrete pieces?

If I accept the limits of my own vision, then I can't see anything other than the clear red acetate. But with an optical microscope I could magnify the die by a factor of fifteen hundred, which would scale it up to the size of a large building. Peering inside this huge cube wouldn't reveal much about how it is built. Everything would still look pretty smooth and continuous.

In the twentieth century, microscopes exploiting the electromagnetic spectrum allowed scientists to magnify things a further thousand times. Now my die will span from one side of London to the other. At this magnification it starts looking grainier. The sense of continuous structure gives way to something more discrete. Current electron microscopes allow me to zoom in another ten times closer, at which point I might start to see the carbon and oxygen atoms that I know to be ingredients of the acetate from which my die is made.

Scientists formulated an atomic view of matter long before they could actually see atoms under a microscope. (The word *atom* was coined by the ancient Greeks to denote the indivisible particles from which all matter was built.) But atoms like oxygen and carbon turned out not to be as indivisible as their name suggested. Now we know that atoms are made up of electrons, protons, and neutrons, which in turn are made up of quarks. In 2013, quantum microscopes captured pictures of electrons orbiting the nucleus of a hydrogen atom. But is there a limit to how far we can dig?

What will happen if I take my die and keep dividing it in half? The mathematical side of me says no problem. If I have a number, I can keep dividing it by two.

$$1, \ \frac{1}{2}, \ \frac{1}{4}, \ \frac{1}{8}, \ \frac{1}{16}, \ \ldots$$

There is no point, mathematically, at which I have to stop. But if I try to do the same thing with the physical die sitting on my desk—if I cut it in half, then in half again—how far can I keep going? The tension between the continuous versus the discrete nature of matter, between what is mathematically possible and what is physically possible, has been raging for millennia. Is the universe dancing to the sound of my trumpet or shimmying to the glissando of my cello?

THE MUSIC OF THE SPHERES

How do I know about the electrons and quarks that are believed to be the last layer of my die? I've never seen them. I've actually forgotten when or why I first became aware of their existence. It is a bit like the way I know that Everest is the tallest mountain. I know it only because I've been told so many times. But before asking whether there is anything beyond electrons and quarks, I want to know how we got to these basic building blocks in the first place.

It's only just over a hundred years ago that convincing evidence emerged that things are made up of discrete building blocks, but the hunch that this was the case goes back thousands of years. In India it was believed that matter was made of basic units corresponding to taste, smell, color, and touch. The Indians divided these units into some that were infinitesimally small and took up no space, and others that were "gross" and took up finite space—an extremely prescient theory, as you will see once I explain our current model of matter.

In the West the ancient Greeks were the first to propose an atomistic philosophy of nature, advocating the view that physical reality could be reduced to fundamental units that made up all matter and could not be broken down into anything smaller. One of the seeds for this belief was a rather remarkable discovery attributed to Pythagoras: namely, that numbers form the basis of musical harmony. Legend has it that inspiration struck when he passed a blacksmith and heard hammers banging out a combination of harmonious notes. Intrigued by the musical harmony, he went home and experimented with the notes made by a stringed instrument.

If I take the vibrating string on my cello, I can produce a continuous sequence of notes by gradually pushing my finger up toward the bridge of the cello, making a sound called a glissando. If I stop at the positions that sound harmonious when combined with the open vibrating string, it turns out that the lengths of the strings are in a perfect whole-number ratio with each other. If I place my finger at the halfway point along the vibrating string I get a note that sounds almost exactly like the note I started with. The interval is called an octave, and to the human ear the notes sound so similar that we give them the same names. If I place my finger a third of the distance from the head of the cello, I get a note that sounds particularly harmonious when combined with the note of the open string, known as the perfect fifth. Our brains respond to a subliminal recognition of this whole-number relationship between the wavelengths of the two notes.

Having found that whole numbers were at the heart of harmony, the Pythagoreans began to build a model of the universe with these whole numbers as the fundamental building blocks of everything they saw or heard around them. Greek cosmology was dominated by the idea of a mathematical harmony in the skies. The orbits of the planets were believed to be in a perfect mathematical relationship to each other, giving rise to the idea of the music of the spheres.

More importantly for understanding the makeup of my die, the Pythagoreans also believed that discrete numbers were the key to understanding the physical world. They were the first to propose the idea that atoms, like numbers, could be added together to make up matter. Plato took this one step further and argued that these atoms were shapes, which he believed to be the key to the ingredients of Greek chemistry: the elements of fire, earth, air, and water. Each element, Plato believed, had its own three-dimensional mathematical shape. Fire was the shape of a triangular-based pyramid, or tetrahedron, made from four equilateral triangles. Earth was cube-shaped, like my Vegas dice. Air was made from a shape called an octahedron, constructed from eight equilateral triangles. It is a shape that looks like two square-based pyramids fused together along the square faces. Finally, water corresponded to the icosahedron, a shape made from twenty equilateral triangles. Plato believed that the geometrical interaction of these basic shapes gave rise to the chemistry of the elements.

An atomistic view of matter was not universally held across the ancient world. After all, there was no evidence for these indivisible bits. You couldn't see them. Aristotle was one of those who did not believe in the idea of fundamental atoms. He thought that the elements were continuous in nature, that you could theoretically keep dividing my die up into smaller and smaller pieces. He believed that fire, earth, air, and water were elemental in the sense that they could not be divided into "bodies different in form." If you kept dividing, you would still get water or air. If you take a glass of water, to the human eye it appears to be a continuous structure that, theoretically, can be infinitely divided. Likewise if you take a piece of rubber, you can stretch it in a smooth manner, making it appear continuous in nature. The stage was set for the battle between the continuous and the discrete models of matter. The glissando versus the discrete notes of the musical scale. The cello versus the trumpet.

Intriguingly, it was a discovery credited to the Pythagoreans that would threaten the atomistic view and turn the tide for many years in favor of the belief that matter could be divided infinitely.

IRRATIONAL EXUBERANCE

If I draw two lines on the page, from an atomistic point of view each line will consist of a certain number of indivisible atoms whose lengths would be in a ratio corresponding to the number of atoms making up each line. But things didn't turn out to be quite so orderly. In fact, it was Pythagoras's own theorem about right-angled triangles that revealed that the world of geometry could give rise to lines whose relative lengths could not be captured by simple fractions.

My casino die offers a challenge to this atomistic view of nature. Take two of the edges of the cube at ninety degrees to each other. They have equal length. Now consider the diagonal line across the face of the cube that completes the triangle. How long is this diagonal line relative to the shorter sides?

Pythagoras's theorem about right-angled triangles says that the square of the length of the diagonal is equal to the sum of the squares of the lengths of the two shorter sides. If I set the length of the sides of my die as 1, then Pythagoras's theorem implies that the length of the diagonal across its face is a number that, when you square it, is equal to 2. So what is this number?

The Babylonians, it turns out, were fascinated by the challenge of solving this riddle. A tablet housed at Yale University dating back to the Old Babylonian period (1800–1600 BC) offers an estimate. Written using the sexagesimal system, or base 60, the Babylonians proposed that the length be:

$$1 + \frac{24}{60} + \frac{51}{60^2} + \frac{10}{60^3} = \frac{30,547}{21,600}$$

In decimal notation this comes out to $1.41421296296296\ldots$, where the 296 repeats itself infinitely. It is true of all fractions that when written as decimals, they repeat themselves after some point (and any decimal expansion that repeats itself can always be written as a fraction). The Babylonian calculation is quite a feat. It is correct up to six decimal places. But when you square that fraction it just misses 2. The ancient Greeks discovered that however hard the Babylonian scribes tried, they would always find that their fractions just missed squaring to 2.

One of Pythagoras's followers, Hippasus, is credited with the discovery that the Babylonians were doomed to failure. He proved that the length of this diagonal could never be expressed as a fraction. Pythagoras's theorem about right-angled triangles implies that the long side would equal the square root of 2 times the length of the short sides. But Hippasus proved that there was no fraction whose square was exactly 2. This length would have to be expressed by a new sort of number. The proof uses one of the classic tools in the mathematician's arsenal: proof by contradiction. Hippasus began by assuming that there was a fraction whose square was 2. By some deft manipulation this always led to the contradictory statement that there was a number that was both odd and even. The only way to resolve this contradiction was to realize that the original assumption must have been false—there can be no fraction whose square is 2.

His fellow Pythagoreans were reputedly dismayed by the revelation that their beautiful right-angled triangles could produce such inharmonic lengths. The sect took a vow of silence. When Hippasus let the discovery out of the bag, legend has it that he was drowned at sea for revealing such disharmony in the physical world. But these new numbers, called irrational numbers because they are not ratios of whole numbers, were not so easily silenced.

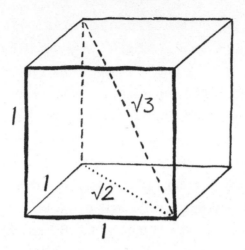

Irrational lengths inside the cube

I certainly have the feeling that this length exists. I can see it on a ruler held up against the long side of the triangle. It is the distance between two opposite corners of any side of my die. Yet if I try to write down the number as an infinite decimal I can never capture it. It begins 1.414213562 . . . and then continues to infinity, never repeating itself.

The discovery by the ancient Greeks that there were lengths that couldn't be expressed as simple ratios of whole numbers led the mathematicians of the time to create new mathematics, the mathematics of irrational numbers, to take the measure of the universe. Other basic lengths like π, the circumference of a circle of unit diameter, also turned out to be irrational and not captured by ratios of whole numbers. Although the ancient Greeks knew two thousand years ago about the irrationality of the square root of 2, it took until the eighteenth century for Swiss mathematician Johann Heinrich Lambert to prove that no fraction could capture π.

Reading about numbers that can't be captured using simple fractions was one of the defining moments that sparked my love affair with mathematics. The same year that my music teacher introduced me to the trumpet in the storeroom cupboard, my math teacher introduced me to the proof of the irrationality of the square root of 2. The proof was contained in a book called *A Mathematician's Apology*, which he recommended to ignite the mathematical fire in me. It worked. I was amazed to find that you could prove with a finite logical argument that only the infinite could articulate a simple length such as the diagonal across a square. If I couldn't write out this length, the next best thing was to know why.

Since reading that proof as a schoolkid, I have learned about alternative ways to explore irrational numbers. There are infinite expressions with patterns that make the number less mysterious. For example,

$$\sqrt{2} = 2 \times \left(1 - \frac{1}{3}\right) \times \left(1 + \frac{1}{5}\right) \times \left(1 - \frac{1}{7}\right) \times \left(1 + \frac{1}{9}\right) \times \cdots$$

or

$$\pi = 4 \times \left(1 - \frac{1}{3} + \frac{1}{5} - \frac{1}{7} + \frac{1}{9} - \cdots\right)$$

The discovery of these expressions pulled irrational numbers into the known. A fraction is a number whose decimal expansion repeats itself from some point, and I comforted myself with the thought that these expressions formed a clear pattern, which helped me to know these new numbers. Of course, for any practical application, I could probably get away with an approximation in the form of a fraction. Most engineers are happy to use the estimate 22/7 for π, which Archimedes got by approximating a circle with a ninety-six-sided figure. In fact, I need only know thirty-nine digits of π to be able to calculate the circumference of a circle the size of the observable universe to the precision of a hydrogen atom.

But only by dividing space infinitely would I be able to measure the precise dimensions of my simple cube. The discovery of irrational numbers meant that Aristotle's continuous view of matter would remain dominant in the West until the Renaissance. Newton's contemporary Robert Boyle appears to have been the first to question the Aristotelian view of matter, which had dominated for nearly two thousand years. In his book *The Skeptical Chymist*, Boyle challenged the idea that matter was made up of the four elements of fire, earth, air, and water. These might be good descriptions of the states of matter, he argued, but not of its constituent parts.

Instead, he argued for the existence of chemical elements. Not only that, he also made what at the time was a rather heretical statement. He believed that these elements were minute bodies differing only in "bulk, figure, texture, and motion." This was theologically dangerous, as it seemed to suggest a materialistic and godless view of the world. Because of this, some have gone as far as to declare Boyle the Galileo of the chemical revolution.

Although Newton would concur with Boyle's suggestion that the material world was made up of indivisible units, the mathematical tools he developed at the same time relied heavily on the notion that time and space were infinitely divisible. Calculus, which allowed one to take a snapshot of a universe in constant flux, made sense only as a process in which space is divided up into ever smaller pieces.

The question of the infinite divisibility of time and space had fueled philosophical argument since the days of the ancient Greek thinker Zeno of Elea. In one of his famous paradoxes, Zeno proposed that an arrow can never hit its target because it must first cover half the distance to the target, then half the distance again, and then half again, requiring infinitely many moves if it is ever going to make it to the target.

The success of Newton's calculus reignited the debate. There were still those who considered the notion of infinite divisibility almost heretical. One such, Bishop Berkeley, dedicated a whole treatise, *The Analyst,* to arguing that trying to make sense of dividing by zero was absurd. The subtitle made his intentions clear—it was "Addressed to an Infidel MATHEMATICIAN."

While most infidel mathematicians quickly latched onto the power of calculus, Newton's other breakthroughs supported the view that, although space and time could be infinitely divided, matter could not. His idea of a world made up of indivisible matter would in time become the prevalent theory of the universe. But at this point, it was still very much a theory without a great deal of evidence.

Newton's discovery of the laws of motion—of how forces acted on objects such as planets and apples—had been so successful that Newton came to believe that if it held true for the very large and the medium sized, then why not for the very small? Maybe all matter consisted of tiny particles whose behavior was determined by his laws of motion. In his *Principia* he stated his belief that by applying his ideas to these particles one could predict the behavior of all material things.

Newton's theory of light also contributed to the growing feeling that an atomistic perspective was the best way to understand the world. The easiest way to describe the phenomena he documented in his book *Opticks* was to think of light as a particle. The way it reflected seemed to mimic the behavior of a billiard ball as it bounced off the sides of the table. From a scientific point of view, there was no empirical evidence for this model of a universe made from indivisible particles. Even with the new microscopes that began to emerge in the seventeenth century, you couldn't see anything to justify an atomistic model of matter. But atoms began to permeate the popular culture of the time. Nicholas Brady's "Ode to Saint Cecilia," set to music by Purcell in 1691, talks of seeds of matter:

> Soul of the World! Inspir'd by thee,
> The jarring Seeds of Matter did agree,
> Thou didst the scatter'd Atoms bind,
> Which, by thy Laws of true proportion join'd,
> Made up of various Parts one perfect Harmony.

The best evidence for an atomistic view of matter came a hundred years later, and it was, as Brady had intimated, "made up of various Parts one perfect Harmony."

ATOMIC ALGEBRA

At the beginning of the nineteenth century, English chemist John Dalton provided the first real experimental justification for the notion that matter was made up of indivisible atoms. His discovery that compounds were made up of substances that were combined in fixed whole-number ratios was a breakthrough, and it led to a new scientific consensus that the building blocks of matter really did come in discrete packages.

"The elements of oxygen may combine with a certain portion of nitrous gas or with twice that portion, but with no intermediate quantity," he wrote. This certainly didn't constitute a proof that matter was discrete, and it was not strong enough evidence to knock the belief of those who favored a continuous model of matter. But it was highly suggestive.

The combination of nitrogen and oxygen could be expressed algebraically as $N + O$ or $N + 2O$. There was nothing in between. All compounds seemed to come in whole-number ratios. For example, aluminum sulfide was given algebraically as $2Al + 3S = Al_2S_3$. In this instance the elements combined in a 2:3 ratio. Elements never combined in a non-whole-number relationship. It was as if a musical harmony existed at the heart of the chemical world. The music of tiny spheres.

The Russian scientist Dmitri Mendeleev is remembered for laying out this growing list of molecular ingredients in such a way that a pattern began to emerge. Like several scientists before him, Mendeleev arranged the elements in increasing relative weight, but he realized that to get the patterns he could see emerging he needed to be flexible. He'd written the known elements down on cards and was continually moving them around on his desk in a game of chemical patience, trying to get them to yield their secrets. But nothing worked. It was driving him crazy. Eventually he collapsed in exhaustion and the secret emerged in a dream. When he awoke, his dream gave him the pattern for laying out the cards. One of the important points that led to his successful arrangement was the realization that he needed to leave some gaps—that some of the cards from the pack were missing.

The key to his arrangement was something called the atomic number, which depended on the number of protons inside the nucleus rather than the combination of protons and neutrons that determines the overall weight. Since no one had any clue yet about these smaller particles, Mendeleev was guessing at its underlying reason. Next he noticed that elements eight along seemed to share very similar properties. Eight on from lithium was sodium, followed after another eight by potassium. All of these were soft, shiny, highly reactive metals. Similar patterns matched up gases with related properties.

This rule of eight had been picked up before Mendeleev's breakthrough and was called the law of octaves. It was compared to the musical octave: if I play the eight notes of a major scale on my cello, the top and bottom notes sound very similar and are given the same letter names. When John Newlands, the originator of this law of atomic octaves, first proposed his concept, he was laughed out of the Royal Society. "Next you'll be trying to tell us that the elements can be understood by putting them in alphabetical order," joked one Fellow. Mendeleev's arrangement confirmed to a certain extent the veracity of this law of octaves. This idea of repeating or periodic patterns led to Mendeleev's arrangement being called the "periodic table."

Mendeleev's genius was to realize that if the elements didn't quite match up, it might simply indicate a missing element that had yet to be discovered. The gaps in his table were probably his most insightful contribution. The fact that there was a hole in the thirty-first place, for example, led Mendeleev to predict in 1871 the existence and properties of a new substance that would later be called gallium. Four years later, French chemist Lecoq de Boisbaudran isolated the first samples of this new element, predicted thanks to the mathematical patterns discovered by Mendeleev.

Here was a list of the elements that made up all matter. My die, for example, is made by putting together carbon atoms, oxygen atoms, and hydrogen atoms in a structure called cellulose acetate (my own body is predominantly made up of combinations of these same atoms but with a different structure). Cellulose acetate is a homogeneous structure, free of bubbles. Older dice were made from a nitrate-based cellulose concocted by John Wesley Hyatt in 1868. His cocktail of nitric acid, sulphuric acid, cotton fibers, and camphor produced an impressive substance with great tensile strength that resisted the effects of water, oils, and even diluted acids.

Hyatt's brother named their new substance "celluloid," and it became a highly cost-effective substitute for objects that had previously been carved out of ivory or horn. Billiard balls, removable collars, and piano keys were all made from this synthetic plastic. Dice made from cellulose nitrate were the industry standard in the early twentieth century, but after several decades of use they would crystallize and decompose, crumbling in on themselves and releasing nitric acid gas. The real collectors' items are Vegas dice made from cellulose nitrate in the late '40s that avoided this crystallization. Mine won't suffer the same fate. On the next page is a picture of how its atoms are put together.

The identification of these elements was not proof of a discrete model of matter. There was no reason why this pictogram couldn't be a formula for the way a continuous structure combines. Although chemists were tending toward an atomistic view of the universe, this was far from the case among physicists. The few who embraced an atomic model of matter were laughed out of the lab.

Ludwig Boltzmann, a German physicist, challenged his peers. He believed that atomic theory offered a powerful way to explain the concept of heat. He argued that a gas was made up of tiny molecules bashing around like a huge game of micro-billiards. Heat was just the combined kinetic energy of these tiny moving balls. Using this model, combined with ideas of probability and statistics, he was successfully able to explain the large-scale behavior of a gas. Most physicists, committed as they were to a continuous view of matter, dismissed his ideas.

Boltzmann was so fiercely ridiculed that he was forced to retreat from his belief that this billiard-ball theory of matter represented reality, and to refer to it as a heuristic model if he wanted to get his ideas in print. Ernst Mach, his great nemesis, taunted him, saying: "Have you ever seen an atom?" Boltzmann was plagued by depression (there is evidence that he was, in fact, bipolar). The rejection of his ideas by the scientific community contributed to the depression that struck in 1906. That summer, he hung himself during a holiday with his family near Trieste, while his daughter and wife were out swimming.

This was a tragic end, not least because evidence that he was right was just emerging. The work that Einstein and others did on Brownian motion would prove extremely difficult to explain for those, like Mach, who believed in a continuous view of the world.

POLLEN PING-PONG

Although conventional microscopes don't allow one to see individual atoms, nineteenth-century microscopes did allow scientists to see the effect these atoms were having on their surroundings. In 1827, Robert Brown was struck by the random behavior of small particles of pollen floating on the surface of water. Since pollen was organic, Brown's first thought was that it might be exhibiting signs of life as it jumped around the surface. When Brown saw the pollen's behavior replicated by inorganic matter like coal dust, he was rather stumped as to what was causing the jittery motion.

The idea that invisible atoms might be bashing around in such a way as to impact the visible world had first been suggested by the Roman poet Lucretius in his didactic poem *On the Nature of Things*:

> Observe what happens when sunbeams are admitted into a building and shed light on its shadowy places. You will see a multitude of tiny particles mingling in a multitude of ways . . . their dancing is an actual indication of underlying movements of matter that are hidden from our sight. . . . It originates with the atoms which move of themselves. Then those small compound bodies that are least removed from the impetus of the atoms are set in motion by the impact of their invisible blows and in turn cannon against slightly larger bodies. So the movement mounts up from the atoms and gradually emerges to the level of our senses, so that those bodies are in motion that we see in sunbeams, moved by blows that remain invisible.

This was written in 60 BC, but it would take Einstein's mathematical analysis of the motion to confirm this atomic explanation of the random movement in Lucretius's sunbeams and Brown's pollen. Nowadays we call it Brownian motion.

Could some model explain the strange motion Brown had observed exhibited by pollen dancing on the surface of the water? If you divide the surface into a grid, there seems to be an equal probability that the pollen will move left, right, up, or down. Its motion is similar to that of a drunken man who randomly stumbles this way and that. The picture below shows the paths of various particles of pollen as plotted by the French physicist Jean Baptiste Perrin, who took up the challenge of explaining the pollen's motion in his book *Les Atomes*.

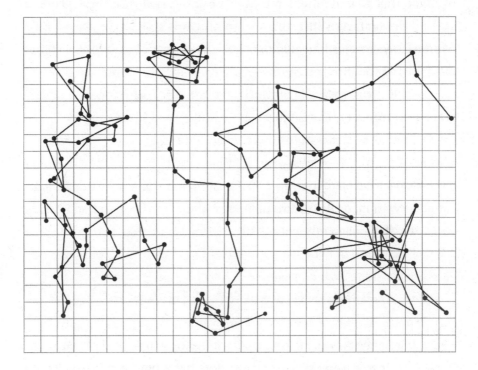

It was only at the beginning of the twentieth century that scientists were finally able to prove that the pollen was being buffeted by the motion of much smaller molecules of water.

Thanks to his mathematical brilliance, Einstein came up with a model that explained how a large object was subjected to the impact of much smaller objects moving randomly. He proved that his model predicted precisely the observed behavior. Think of an ice rink with a large puck sitting in the middle of the rink and then introduce a whole system of tiny pucks that are set off in random directions at particular speeds. Every now and then the tiny pucks will hit the large puck, causing it to move in one direction. The challenge was to assess how many small pucks you would need, and their relative size, in order to produce the observed behavior of the larger puck.

Einstein's success in producing a mathematical model that replicated the motion of the pollen was a devastating blow to anyone who believed that a liquid like water was a continuous substance. No one still believing in Aristotle's view of matter was able to come up with a comparably convincing explanation. And yet, although Einstein's model offered convincing evidence that matter came in discrete pieces, it did not answer the more fundamental question of whether you could infinitely divide these pieces into ever-smaller parts.

Indeed, the indivisible atoms turned out to be far from indivisible. Scientists would soon discover smaller constituent parts. The next layer down revealed that an atom is made up of even tinier pucks called electrons, protons, and neutrons, the first of which had already come to light some eight years before Einstein's breakthrough.

PULLING APART THE ATOM

The way science works is that you hang on to your model of the universe until something pops up that doesn't seem to fit—something new that you can't quite explain with the current model. Sometimes the challenge comes from an altogether different field. This was the case with the discovery of subatomic particles.

The realization that the atom might be made up of smaller bits emerged from experiments British physicist J. J. Thomson conducted at the end of the nineteenth century to understand electricity. Thomson had been investigating how gas conducted electricity. His early experiments involved a single glass tube filled with air with two electrodes at either end. By applying a high-voltage charge between the electrodes, an electric current was produced. The strange thing was that Thomson seemed to be able to actually see the current, as an arc of light that appeared in the air between the two electrodes.

Things became even stranger when he removed the gas completely from the tube and applied the voltage across a vacuum. The arc of light disappeared. But, bizarrely, the glass at the end of one tube was found to fluoresce. When he placed a metal cross in the tube, a cross-shaped shadow appeared in the middle of the glowing fluorescent patch.

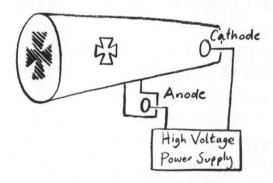

Electrons emitted from the cathode that hit the
opposite wall cause the glass to fluoresce.

The shadow always appeared opposite the negative electrode, otherwise known as the cathode. The best explanation was that the cathode was emitting some sort of ray that interacted with matter and made either the gas in the tube or, in the case of the vacuum, the glass of the tube itself glow.

These "cathode rays" were something of a mystery. They were found to pass right through thin sheets of gold when they were placed in the way. Were they some sort of wave-like phenomenon, like light? Some scientists thought they were made up of negatively charged particles spat out by the negative electrode and then attracted to the positive electrode. But how could these particles pass through solid gold?

If they were negatively charged particles, Thomson believed he should be able to change their path through the tube by applying a magnetic field. The German physicist Heinrich Hertz had already tried this and failed, but Hertz hadn't removed enough gas, which interfered with the experiment. With the gas fully removed, things worked just as Thomson had hoped. Apply a magnetic field to the rays, and, sure enough, the shadow shifted. The rays were being bent by the magnet.

The real surprise came when Thomson made a mathematical calculation of what the mass of these charged particles must be. If you apply a force to a mass then, as Newton's laws of motion state, the amount you'll be able to move it will depend on the mass. So the amount of deflection that a magnetic field will cause will have encoded in it information about the mass of these particles. The calculation also depends on the charge. Once this was determined in a separate experiment, Thomson could finally work out the mass of his mysterious charged particles. The answer was startling. It was nearly two thousand times smaller than the mass of a hydrogen atom, the smallest atom in the periodic table.

That these particles seemed to originate from the metal making up the electrode led to the hunch that these particles were actually smaller constituents of the atom. The atom wasn't indivisible after all: there were smaller bits. They were called electrons, from the Greek word for amber, the first substance to exhibit a charge.

The discovery that atoms were made up of even smaller constituents was a shock to many scientists' view of the world. When Thomson gave a lecture on his findings, he was met by startled disbelief. As he wrote with his characteristic wit, "I was told long afterward by a distinguished physicist who had been present at my lecture that he thought I had been pulling their leg."

When Thomson used a different metal, the masses of particles didn't change. This seemed to suggest that every atom had the same microparticles as constituent parts. The first thought was that a hydrogen atom, given that it is two thousand times heavier than this new electron, might be made up of two thousand or so of these electrons. But a helium atom was roughly twice the mass of a hydrogen atom. Why would the number of electrons jump from two thousand to four thousand, with nothing in between? Whole-number ratios between the masses of atoms in the periodic table had been one of the reasons for supposing that they were truly atomic. So what could account for these discrete steps in mass? Furthermore, atoms were electrically neutral. So were there other particles that canceled out the charge of the electron? Could you get atoms to emit positive particles to counter these negative electrons?

The experiments offered evidence of a positive ray of particles running in the opposite direction. When a magnetic field was applied to these particles, they were much harder to deflect, implying that they were larger than the electrons. But their masses seemed to vary according to the gas that was being used to fill the tubes. Thomson started with hydrogen and found that the mass of the particles in this positive ray was essentially the same as the mass of the atom you started with. It seemed that the hydrogen atoms in the tube were having their electrons stripped away, leaving a large positive particle that was then attracted to the opposite electrode.

Thomson managed to achieve a similar effect with other gases such as helium, nitrogen, and oxygen. The masses were all whole-number multiples of the positive particle produced by the hydrogen atom. Atomic harmony yet again. As yet, there was no reason to believe that there weren't many sorts of positive particles, just as there were many sorts of atoms. Thomson put forward a model of the atom known as the plum pudding. The positively charged part of the atom formed the pudding making up the bulk of the atom, while the negative electrons were the tiny plums inside.

STRANGE BALLISTICS

Ernest Rutherford is generally credited with the discovery of the proton, the building block for all these positive particles that Thomson had investigated. Just as with the electron, the discovery came in a rather roundabout manner. Rather than the electricity that had interested Thomson, Rutherford was fascinated by the new science of radioactivity. He found that uranium spat out positively charged particles, known as alpha particles, and became interested in what would happen when these particles collided with other atoms. The age of the bombardment of the atom had begun. It would eventually lead to the ultimate atom smasher: the Large Hadron Collider at CERN.

When Rutherford's student Hans Geiger placed a thin sheet of gold foil between a stream of alpha particles and the plate detecting the particles, to Rutherford's surprise the resulting behavior contradicted the theoretical model of the atom. In Thomson's plum pudding model of the atom, which has a positive charge evenly distributed, positive alpha particles passing through the metal would be repelled by the positive charge in the atom. Given that the charge is distributed over the full extent of the atom, you wouldn't expect much deflection.

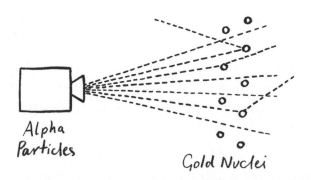

Alpha
Particles

Gold Nuclei

Alpha particles being deflected by the nuclei of atoms of gold

SECOND EDGE: MATTER

But Geiger found that some of the alpha particles were deflected wildly, bouncing back off the gold foil in the direction they'd been fired from. Rutherford was staggered: "It was as if you fired a fifteen-inch shell at a piece of tissue paper and it came back and hit you." By counting how many alpha particles were deflected, and by how much, Geiger and Rutherford discovered that the data were consistent with the charge and mass being concentrated in a tiny center of the atom, which became known as the nucleus. It still wasn't clear whether this nucleus was indivisible.

When Rutherford bombarded lighter atoms with alpha particles, evidence emerged that the nucleus wasn't a single entity and that it was made up of constituent particles. Tracing the paths of the alpha particles in a cloud chamber, he detected paths that were four times longer than they should be. It was as if another particle four times as light was being kicked out of the nucleus by the impact of the alpha particles. Rutherford found that once the particle was kicked out of the nucleus, the atoms transformed. Pure nitrogen was being converted into oxygen by the impact. Knock out one of these particles and the element changed.

Here was evidence for a building block from which all nuclei of atoms were built. It behaved just like the hydrogen atom with its electron stripped off. Rutherford had discovered the proton. The nuclei of atoms were built by taking multiples of this proton. The only trouble was that the charge didn't make sense. The nucleus of helium was four times as heavy as that of the hydrogen atom, yet its charge was only twice as big. How could this be? Perhaps there were electrons in the nuclei attached to protons, canceling out the charge. But electrons and protons couldn't operate in such close proximity, so that couldn't be the answer.

This led Rutherford in the 1920s to guess that there might be a third constituent, which he called a neutron, with approximately the same mass as the proton but no charge. Producing evidence for this

97

particle proved very tricky. He proposed to his colleague James Chadwick crazy ways by which they might reveal the neutron. Experiments conducted in the 1930s in Germany and France eventually picked up evidence of neutral particles emitted when various nuclei were bombarded with alpha particles, but the experimenters mistakenly believed it was some sort of electromagnetic radiation.

Chadwick was convinced that these particles must be the neutrons he'd discussed with Rutherford. Further experiments revealed that their mass was only slightly bigger than the proton, and without charge this new particle was the missing ingredient that made sense of the numbers. With Chadwick's discovery it seemed as if the building blocks of matter had been revealed.

It was an attractive model. Fire, earth, air, and water—Aristotle's four elements—had been reduced to three particles: the electron, proton, and neutron. With these three building blocks, scientists believed they could build all matter. Oxygen: eight protons, eight neutrons, and eight electrons. Sodium: eleven protons, twelve neutrons, and eleven electrons. All matter seemed to be made up of whole-number combinations of these three particles, and for many decades, that is where things stood.

Except that the dividing didn't stop there. It turned out protons and neutrons were not indivisible, but their building blocks—quarks—have a strange property: they don't like to be seen in isolation. They only come in groups. Safety in numbers. If they have never been seen on their own, why are scientists so confident that there are smaller bits into which protons and neutrons can be divided? And if protons and neutrons can be divided, why couldn't these new constituent parts be equally divisible? Perhaps the question of the indivisibility of matter is fundamentally unknowable. The discovery of the next layer, as we probe even deeper into matter, is the topic of our next excursion.

4

Everything we call real is made of things that cannot be regarded as real.

—Niels Bohr

T HE REALIZATION THAT protons and neutrons might not be indivisible came not from new and more sophisticated technology, but from the mathematics of symmetry. It is striking that time and again, mathematics appears to be the best lens we have to peer into the mysteries of science. The mathematical model responsible for our belief in the divisibility of the proton and neutron arose shortly after physicists discovered that there were many more particles out there than the three that were believed to be the constituent parts of stable atoms. The discovery of these new particles emerged from the study of collisions that happen in the upper atmosphere when cosmic rays strike the atmosphere.

The first evidence of new particles was found in the cloud chambers that scientists had built in their labs to record the path of charged particles. (Cloud chambers are sealed tanks full of a supersaturated vapor of water and alcohol; the supersaturation is such that any charged particle passing through it leaves a trail of condensation.) Carl Anderson, a physicist working at the California Institute of Technology, had used cloud chambers in 1933 to confirm the existence of a strange new thing called antimatter, first predicted some years earlier by British physicist Paul Dirac. Dirac's attempt to unify quantum physics and the theory of electromagnetism had successfully explained many things about electrons, but it had also suggested something that didn't correspond to anything anyone had seen in the lab.

Dirac's equations were a bit like the equation $x^2 = 4$. The obvious solution is $x = 2$, but there is also another mirror solution, namely $x = -2$, because -2×-2 is also equal to 4. The mirror solution in Dirac's equations suggested that there might be a mirror version of the electron with a positive charge. Most thought this was a mathematical curiosity with no basis in reality, but when, four years later, Anderson spotted in his cloud chamber traces of a particle behaving like an electron in a mirror, antimatter went from theory to reality. Anderson's "positrons," as they came to be known, could be found in the particle interactions happening in the upper atmosphere. And they weren't the only new things to appear.

Even stranger particles were soon leaving trails in Anderson's cloud chamber. Anderson started to analyze these unanticipated negatively charged new particles with his doctoral student Seth Neddermeyer in 1936. The paths they were leaving indicated that their mass was much larger than that of an electron (mass can be measured by how much a particle is deflected under the influence of a magnetic field). The particle seemed to have the same charge as the electron, but it was much harder to deflect.

The muon, as this particle is now called, was one of the first new particles to be discovered in the interactions of cosmic rays with the atmosphere. The muon is unstable. It quickly falls apart, most often into an electron and a couple of neutrinos (another new particle, first detected in the 1950s) with almost no mass and no charge. Neutrinos made sense both of neutron decay and of the decay of this new muon. The decay rate of the muon was on average 2.2 microseconds, slow enough for some particles to still be around when they reached the surface of the Earth.

The muon helped confirm Einstein's prediction that time slows down as you approach the speed of light. Given its half-life, far fewer muons should be reaching the surface of the Earth than were being detected. The fact that time slows down close to the speed of light helped explain this discrepancy. I will return to this in the Fifth Edge, when we consider pushing time to the limits of knowledge.

The muon behaved remarkably like the electron, but it had greater mass and was less stable. When the American physicist Isidor Rabi was told of the discovery, he said, "Who ordered that?" It seemed strangely unnecessary for nature to introduce a heavier, less stable version of the electron. Little did Rabi realize how much more there was on the menu.

Having realized that the interactions of cosmic rays with the upper atmosphere were creating new forms of matter, physicists decided that they had better not wait for these particles to reach the cloud chambers in their labs, by which time they might have decayed into traditional forms of matter. So the cloud chambers were moved to high-altitude locations in the hope of picking up other particles.

The Caltech team chose the top of Mount Wilson, near their home base in Pasadena. Sure enough, new tracks indicated that new particles were being picked up. Other teams placed photographic plates in observatories in the Pyrenees and the Andes to see if they could record different interactions. It turned out that the muon was the least of Rabi's worries. A whole menagerie of new particles started showing up.

Some had masses about one-eighth that of a proton or neutron. They came in positively or negatively charged varieties and were dubbed pions. An electrically neutral version, harder to detect, was later discovered. In Manchester two photographs showed what appeared to be a neutral particle decaying into pions. The mass of these new particles was roughly half that of a proton. The cloud chamber at the top of Mount Wilson recorded more evidence to support the discovery of what would soon become known as kaons.

As time went on, more and more particles were uncovered, so

many that the whole thing became totally unwieldy. As Nobel Prize winner Willis Lamb quipped in his acceptance speech of 1955, "The finder of a new elementary particle used to be rewarded by a Nobel Prize, but such a discovery now ought to be punished by a $10,000 fine." It had been hoped that the periodic table would be simplified once scientists had discovered how it was put together using electrons, protons, and neutrons. But these three particles turned out to be just the tip of the iceberg. Now there were more than a hundred particles that seemed to make up the building blocks of matter. Enrico Fermi went so far as to tell a student at the time, "Young man, if I could remember the name of these particles, I would have been a botanist."

Just as Mendeleev had managed to find some sort of order with which to classify and make sense of the atoms in the periodic table, the search was on for a unifying principle that would explain these new muons, pions, kaons, and numerous other particles. The underlying structure that finally made sense of this menagerie of particles— the map, as it were, to the zoo—was ultimately discovered thanks to nimble mathematics.

MAPPING THE PARTICLE ZOO

When you are trying to classify things, whether plants, animals, or subatomic particles, it helps to recognize their dominant characteristics. In the case of animals, the concept of species offers some order in the animal kingdom. In particle physics, one of the characteristics that helped divide the zoo into smaller groups was the idea of charge. How does the particle interact with an electromagnetic force? Electrons would bend one way, protons the other, and the neutron would be unaffected. As new particles emerged, they could be filtered through the gateway of electromagnetic force. Some would join the electron's cage, others would head toward the proton, and the rest would be put together with the neutron—a first pass at imposing some order on our particle menagerie.

Electromagnetic force is one of four fundamental forces identified as being at work in bringing the universe together. The other forces are gravity; the strong nuclear force responsible for binding protons and neutrons together at close quarters; and the weak nuclear force, which controls things like radioactive decay.

What other characteristics could distinguish the different behaviors of these new particles? It was soon found that the mass of a particle was actually quite a good way of establishing some hierarchy. Pions and kaons were a factor lighter than the protons or neutrons that made up ordinary matter, and could thus be collected together. But mass was a rather rough and ready way of sorting particles. Physicists were on the lookout for something more fundamental—a pattern as effective as the one Mendeleev had discovered with his periodic table.

The key to making sense of the onslaught of new particles was a new property called strangeness. The name arose due to the unusual behavior demonstrated by some of these new particles as they decayed. Because mass is equivalent to energy (this is the crux of Einstein's equation $E=mc^2$), and nature favors low-energy states, particles with larger mass often try to find ways to decay into particles with smaller mass.

There are several mechanisms for this decay, each of which depends on one of the four fundamental forces. Each mechanism has a characteristic signature, helping physicists to understand which fundamental force is causing the decay. Energy considerations determine the most likely force at work in the decay of any given particle. The strong nuclear force is usually the first to have a go at decaying a particle, and this will generally decay the particle within 10^{-24} of a second. Next in the hierarchy is electromagnetic force, which might result in the emission of photons. The weak nuclear force is the most costly in energy terms, and so it takes longer. A particle that decays via the weak nuclear force is likely to take 10^{-11} seconds to decay. That's a factor of 10 trillion times slower. So by observing the time a particle takes to decay, scientists can get some indication of which force is at work.

Imagine a ball sitting in a valley flanked by two hills of differing sizes. There is a path to the right that, with a little push, will take the ball over the lower hill into a lower valley. This path corresponds to the strong nuclear force. To the left is a higher hill that also offers a path to the lower valley. This direction represents the work of the weak nuclear force.

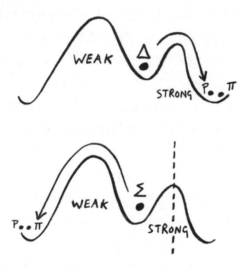

A Delta baryon Δ decays via the strong nuclear force to a proton and a pion. In contrast, a Sigma baryon Σ decays via the weak nuclear force.

So why do some particles find a way over the easy hill while others go the longer way? This seemed rather strange. Certain particles appeared to encounter a barrier (represented in the figure by a broken line) that prevented them from crossing via the easy route to the lower valley.

The physicists Abraham Pais, Murray Gell-Mann, and Kazuhiko Nishijima came up with a cunning strategy to solve this puzzle. They proposed a new property that mediated the way particles interacted with the strong nuclear force. This new property, called strangeness, gave physicists a new way to classify all these new particles. Each new particle was given a measure of strangeness according to whether it would have to take the long route to decay. The idea is that the strong nuclear force can't change the strangeness of a particle, so if you have two particles with different strangeness, the strong nuclear force can't decay one into another. But the weak nuclear force can change strangeness.

Then even more exotic particles were detected, created by higher-energy collisions that seemed to decay in two steps via the weak nuclear force. They were called cascade particles. The proposal was that these were doubly strange, so they were given strangeness number –2. The first step in the decay had been given a strangeness of –1 and after decaying again via the weak nuclear force the particles became protons and neutrons, which had strangeness 0. It may seem a bit like pulling a rabbit out of a hat, but that's part of the process of doing science. You keep pulling stranger and stranger things out of your hat. Most you chuck away because they get you nowhere, but pull enough things out of the hat and every so often you'll get a rabbit. As Gell-Mann later admitted, "The strangeness theory came to me when I was explaining a wrong idea to someone, but then I made a slip of the tongue and I had the strangeness theory." Strangeness turned out to be a pretty amazing rabbit.

Originally the idea of strangeness was meant as a bookkeeping device, something that conveniently kept track of the decay patterns of different particles. It was just another set of cages to help classify the many animals in this particle zoo. But it transpired that this new feature was actually the first hint of a much deeper physical reality at work. The exciting moment came when scientists took particles of a similar mass and began to plot them on a graph measuring their strangeness and charge together. What they got were pictures full of symmetry.

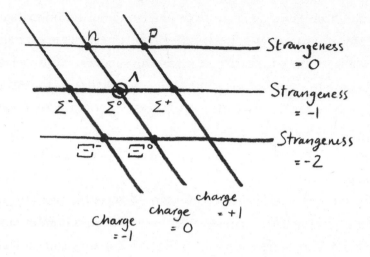

They found themselves with a hexagonal grid, with two particles occupying the center point of the grid. If one took pions and kaons and arranged them on a grid of strangeness against charge, a similar picture emerged. When you get a pattern like this you know you are on to something. The hexagonal patterns these particle pictures were making were not new—they'd been seen before. Not in physics, but in the mathematics of symmetry.

STRANGE SYMMETRIES

To someone trained in the mathematics of symmetry, these arrangements of cages in a hexagonal pattern with a double point at the center look very familiar. They are the signature of a very particular symmetrical object called SU(3). For me this is brilliant. Symmetry is something I know. It gives me a chance to get a handle on what is going on just when things are getting complicated. SU(3) stands for "the special unitary group in dimension 3," but it can describe the symmetries of a range of different geometric objects in different dimensions. Although the objects can look very different, even existing in different dimensional spaces, their underlying symmetries can be the same. The hexagonal grids created by the particles are the same as the images mathematicians use to describe the way SU(3) acts on an object in eight-dimensional space. The eight particles of the hexagonal grid correspond to the number of dimensions you need to create this symmetrical object.

This hexagonal picture was the Rosetta Stone that opened up a whole new world in particle physics. The picture led physicists to the realization that the symmetry group SU(3) was the secret to understanding all these new particles. This new insight became known as "the eightfold way," an invocation of the Buddhist idea of the eightfold way to spiritual enlightenment. Similar pictures, it was soon discovered, could be used to gather together other members of the particle menagerie. Each new picture corresponded to another object whose symmetries were described by SU(3). It was an extraordinary discovery. The fundamental particles were far from being a wild mess. They all seemed to be elements of geometries with a common symmetry.

It is amazing to me how, time and again, the physical world seems to be guided by mathematics. To what extent is it just a good story that helps us make sense of things? Is the physical universe actually a form of physicalized math? This is something I've been asking myself for a long time, and my journey thus far seems to suggest it might well be the case.

Heisenberg was right when he wrote, "Modern physics has definitely decided in favor of Plato. In fact, the smallest units of matter are not physical objects in the ordinary sense; they are forms, ideas which can be expressed unambiguously only in mathematical language." Plato's watery icosahedron and fiery tetrahedron have been replaced by the strange new symmetrical shape of SU(3).

When the physical world turns into a piece of mathematics, I immediately feel that this is something I can comprehend. The mathematics of symmetry is my language. For most people, turning fundamental particles into math means a move away from things they know. Comparing particles to billiard balls or waves gives them a more tangible sense of what is going on. Even the abstract language of eight-dimensional symmetrical objects is possible only because we are abstracting the ideas of things we have physically encountered.

Murray Gell-Mann and Yuval Ne'eman both independently recognized the patterns in these particles in 1961. Ne'eman combined his physics with a career in the Israeli Defense Force, and that year he had been posted to London as a military attaché. He'd meant to study general relativity at King's College, but when he realized that it was miles from the Israeli Embassy in Kensington, he decided to check out what was happening five minutes down the road at Imperial College. It wasn't long before he turned his attention from the very big to the very small.

Although the pattern for the lambda, sigma, and xi particles matched the eight-dimensional symmetry of SU(3), the corresponding pattern of kaons and pions was missing a particle at its center. It was either wrong or there was a new particle to be discovered. Gell-Mann published his prediction of a missing particle in a Caltech preprint in early 1961. Sure enough, physicists in Berkeley discovered the eta particle a few months later.

This is the perfect scenario for a new theory. If you make a prediction that is subsequently confirmed, you know you're on to a winner. In June 1962, Gell-Mann and Ne'eman attended a conference together at CERN. At the conference, a whole host of new particles was announced: three sigma-star baryons with a strangeness of –1, and two xi-star baryons with a strangeness of –2. It was thought that these particles could be used to build a picture that corresponded to another geometric object that had the symmetries of SU(3).

As Gell-Mann and Ne'eman sat there in the lecture, they could both see this new picture emerging, except that one of the corners of the picture was missing. There were only nine particles. The picture should have had ten. Both Gell-Mann and Ne'eman simultaneously realized that one place had been left empty, leading to the prediction of a new particle. Gell-Mann stuck his hand up first and predicted the existence of a new omega particle with a strangeness of –3, a prediction that would be confirmed in January 1964.

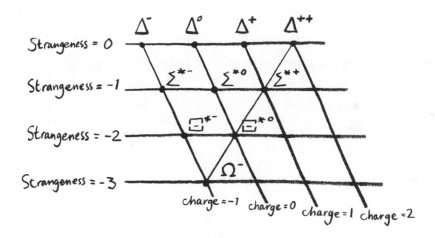

It was a twentieth-century version of Mendeleev's periodic table: the recognition of an underlying pattern with missing jigsaw pieces. Just as the discovery of the missing atoms gave credence to Mendeleev's model, the discovery of these missing particles helped convince scientists that mathematical patterns were powerful ways to navigate the particle menagerie.

The patterns in the periodic table ultimately revealed that atoms were made up of protons, electrons, and neutrons. There was a feeling that the patterns of the newly discovered particles hinted at a similar story—the existence of still more fundamental building blocks.

QUARKS: THE MISSING LAST LAYER?

A number of physicists had noticed that if you place the different pictures that were emerging from all these new particles, you get a pyramid shape—with the top layer missing. If you looked at these layers from the perspective of symmetry, this missing layer, a single triangle, was really the one from which you could grow all the other layers. But no one had spotted any particles that corresponded to this missing layer.

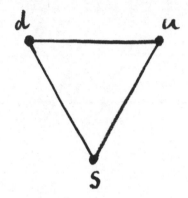

Triangle hinting at three new particles: the up quark *u*,
the down quark *d*, and the strange quark *s*

Robert Serber, Oppenheimer's right-hand man during the Manhattan Project, conjectured that maybe this extra layer suggested the existence of three new fundamental particles. At a lunch with Gell-Mann in 1963, Serber put forward his idea, but when Gell-Mann challenged him to explain what electrical charge these hypothetical particles would have, Serber wasn't sure. Gell-Mann started scribbling on a napkin and soon he had the answer. The charges would be ⅔ or –⅓ of the charge on a proton. The answer seemed ridiculous. "That would be a funny quirk," Gell-Mann commented. Nowhere in physics had anything been observed that wasn't a whole-number multiple of the charge on the electron or proton.

No one had ever seen such a fractional charge. Although Gell-

Mann was initially skeptical about these hypothetical particles with fractional charge, by the evening they began to work their magic on him. In the subsequent weeks he teased out the implications of these ideas, calling the particles "kworks" whenever he talked about them, a word he had used previously to denote "funny little things." Serber believed the word was a play on the idea of the quirkiness Gell-Mann had mentioned at lunch.

It was while perusing James Joyce's experimental novel *Finnegans Wake* that Gell-Mann came across a passage that determined how he would spell the word he was using to describe these hypothetical particles. The opening line of a poem ridiculing King Mark, the cuckolded husband in the Tristan myth, caught his attention: "Three quarks for Muster Mark!"

Given that there were three of these hypothetical new particles that could be used to build the other layers, the reference seemed perfect. The only trouble was that Joyce clearly intended the new word *quark* to rhyme with "Mark" not "kwork." But the spelling and pronunciation Gell-Mann wanted won out.

These quarks would eventually become what we now believe to be the last layer in the construction of matter. But it would take some time for the idea to catch on. During a conversation on the telephone to his former PhD supervisor about these quarks, Gell-Mann was stopped in his tracks. "Murray, let's be serious," his supervisor told him. "This is an international call."

For Gell-Mann, the patterns seemed too beautiful not to have at least some underlying truth to them. He continued to develop his theory that matter was made up of a new layer of three fundamental particles: the up quark, the down quark, and the strange quark, with charge ⅔, –⅓, and –⅓, respectively. All other particles were made up of combinations of these quarks (and their antiparticles in the case of kaons and pions). The number of strange quarks in the makeup determined the strangeness of the particle.

Gell-Mann wasn't the only one to play with the idea of pulling matter apart into these smaller particles. American physicist George Zweig also believed the patterns that had been observed hinted at a more fundamental layer of particles. He called them "aces" and he was even more convinced than Serber or Gell-Mann that these particles had a physical reality. His preprints explaining his ideas were dismissed as "complete rubbish" by the head of the theory group at CERN. Even Gell-Mann didn't initially regard them as more than a mathematical model that created some coherent order in the pictures they were drawing. They were a mnemonic, not concrete reality. Gell-Mann dismissed Zweig's belief in the physical reality of these particles: "The concrete quark model—that's for blockheads."

FROM FANTASY TO REALITY

That would all change when evidence emerged at the end of the 1960s from experiments conducted by physicists working at the Stanford Linear Accelerator Center in which protons were bombarded with electrons. Analysis of a proton's charge revealed that it could be thought to have a size of 10^{-15} meters. The belief was that the proton would be uniformly distributed across this small region. But when electrons were fired at the proton blob, the researchers got a shock from the resulting scatter patterns. The proton, like the atom, turned out to be mostly empty space.

The scattering was, in fact, consistent with a proton that was made up of three smaller particles. Just as with Rutherford's experiments on gold atoms, every now and again one of these electrons would score a direct hit with one of the three points and come shooting back toward the source of the electrons. The experiments seemed to confirm the idea that the proton was made up of three smaller particles. Although a quark has never been seen, the scattering of the electrons indicated, sure enough, that three smaller particles made up the proton.

It turns out the blockheads were right. The up, down, and strange quarks were not just a mathematical mnemonic; they were a physical reality. Over time it was discovered that these three quarks weren't enough to cover all the new particles, and eventually we found ourselves with six quarks together with their antiparticles. In addition to the three Gell-Mann named, three more appeared on the scene: the charm quark, the top quark, and the bottom quark.

The discovery of quarks and their underlying symmetry is one of the most exciting discoveries of the twentieth century. Seeing these fundamental particles lining up in patterns that were already recognized in the mathematics of symmetry must have been so thrilling. If I could choose one discovery in physics I would love to have made, this one would rank pretty high. It must have been like an archeologist coming across a pattern that had previously been seen only in some other distant part of the world, and suddenly realizing there had to be a connection between the two cultures.

The weird thing is that this pyramid of triangles and hexagons giving rise to different representations of SU(3) goes on to infinity, implying that you could keep on gluing together more and more quarks to make more and more exotic particles. Announcements from the Large Hadron Collider in 2015 revealed exciting evidence for a five-quark particle. The researchers at CERN almost missed the particle, called a pentaquark, thinking it was just background noise. But when they tried to remove the noise they discovered a strong signal pointing toward this next layer in the symmetry tower. As one of the researchers working at CERN admitted, "We didn't go out looking for a pentaquark. It's a particle that found us."

How much further can we push our understanding of mathematics to predict what else we might see in the LHC? There is an even bigger symmetrical object called SU(6) that would unite all six quarks—up, down, strange, charm, top, and bottom—into a fusion of fascinating particles. Instead of the two-dimensional pictures scientists have drawn combining particles together in families, you'd need five-dimensional pictures. Although it is possible to cook up some of these more exotic combinations of quarks, the beautiful mathematical symmetry breaks down as the differences in mass between the basic quarks is amplified. Indeed, the top quark is so unstable that it decays before it has time to bind to another quark. Why quarks have different masses seems to be a question that physicists don't know the answer to—yet. The mathematics seems to suggest a much richer cocktail of particles than physical reality can actually sustain. Reality seems a pale shadow of what might be possible mathematically, but understanding that reality still holds many challenges.

I must admit that even with the mathematical toolkit I've spent years assembling, I'm still not sure I know what these quarks really are. I've sat at my desk poring over books on particle physics for months—light fare, such as Sudbury's *Quantum Mechanics and the Particles of Nature*—and lecture notes downloaded for Oxford graduate courses and occasionally begun to despair a little. There is so much that I still don't know: path integrals describing the futures of these particles, the inner workings of the Klein–Gordon equations, what exactly those Feynman diagrams that physicists so easily draw up on the board really mean. I look enviously at my son, who has started his degree in physics. He will have the time to steep himself in this world, to get to know these things as intimately as I know the area I chose to specialize in.

It is the same with my cello. As an adult I hanker to play those Bach suites now, not in ten years' time. But just as it took years to learn the trumpet, so it will be a slow, gradual, sustained period of learning that will bring me to the point where I can play the suites.

As with the cello, I know that if I spend enough time in this world of particle physics I have some hope of understanding what my colleagues across the road in the physics department live and breathe daily. It frightens me to realize I don't have time to know it all. But even those physicists whom I envy for the ease with which they play with our current state of knowledge recognize that ultimately they'll never know for sure whether they know it all.

COWBOYS AND QUARKS

I arranged to meet one of the scientists responsible for discovering one of the last pieces in the quark jigsaw puzzle to see whether particle physicists think there may yet be even tinier pieces. Now a professor at Harvard, Melissa Franklin was part of the team responsible for detecting the top quark at Fermilab in Illinois. Contrary to popular perception, discovering a particle is not a eureka moment but a slow burn. But Franklin told me she preferred it that way: "If it's just 'boom,' it would be a drag. You spend fifteen years building the thing and boom, in one minute it's over? It would be terrible." The team spent almost a full year gathering data before they felt confident enough in 1995 to confirm its discovery.

Franklin is definitely on the experimental rather than theoretical side of the physics divide. Happier with a power drill in her hand than a pencil, she helped to build the detector at Fermilab from the bottom up. We had both been invited to talk at the Rome Science Festival on the theme of the unknowable, so we agreed to meet in the lobby of our rather strange hotel, which seemed to be dedicated to the sport of polo. Given that Franklin strides around the department in cowboy boots, I thought she might feel more at home, but she ended up making an extremely dramatic entrance, crashing to the bottom of the stairs into the lobby. She dusted herself off and strode over as if nothing had happened.

I was keen to know if she thought quarks were the last layer or whether there might be more structure hiding beneath the particles she'd helped to discover.

"We're down to 10^{-18} meters. The next seven or eight orders of magnitude are kind of hard to investigate," she said. "But certainly a lot more could happen in there. It's kind of strange that I could die before—especially if I keep falling down the stairs—I could die before we get any further."

Did she think there might be fundamental limits to what we could know?

"There are definitely limits in my lifetime, but I'm not sure there are any other limits. In experimental physics, saying there's no way we can do something is the perfect way to get someone to figure out how to do it. In my lifetime I'm never going to be able to measure something that decays in 10^{-22} seconds. I don't think there's any way. That's not to say it's provably unknowable.

"We couldn't have imagined the laser or the atomic clock, right? I think all the limitations in physics are going to be atomic because all the things we do are with atoms. I know that sounds weird, but you need the atoms in your detector."

It's intriguing to consider how Einstein inferred the existence of atoms by looking at how they impacted things you could see, like pollen or coal dust. And today we know about quarks because of the way particles bounce off protons. I asked her if she thought there would be new ways to dig deeper.

"I'm sure those guys like Heisenberg and Bohr couldn't have imagined what we can detect today. I guess the same must apply to our generation . . . although of course we're much smarter," she laughed.

I guess that is the problem each generation faces. How can we ever know what cunning new method might be developed to dig deeper into the fabric of the universe? Franklin wondered how much we were missing that was already there in the data coming out of the current generation of detectors.

"Many young people in my field don't believe it's possible to find anything new that wasn't predicted by theorists. That's sad. If you found something and it wasn't predicted by theory, then you'll probably think it's wrong and you'll dismiss it, thinking it's a fluctuation. I worry that because of the way our experiments are constructed you have these triggers that trigger on certain things, but only things you're looking for and not other things. I wonder what we're missing."

That was almost the fate of the pentaquark; it was nearly dismissed as noise. Franklin was intrigued to know whether, if I could press a button and know it all, I would do so. As I was putting out my hand to press her hypothetical button, selling my soul to Mephistopheles to know the proofs of all the theorems I'm working on, she stopped me in my tracks.

"I wouldn't."

"Why not?"

"Because it's not fun that way. There are certain things, like if I could push a button and speak perfect Italian, I would do it. But not with science. I think it's because you can't really understand it that way. You have to struggle with it somehow. You have to actually try and measure things and struggle to understand things."

I was intrigued. Wouldn't she press the button if she could know there were more particles sitting underneath the quarks?

"If they just told me the method, then that would be great. But a lot of the reason we like doing science is coming up with the ideas in the first place. The struggle is more interesting."

I think, in the end, Franklin likes making things. She likes driving forklift trucks and drilling concrete in the search for new particles, and wouldn't want knowledge to come too easily.

"Experimentalists are a bit like cowboys in a way. Lasso that old thing over there and bring it over here. Don't mind that boy over there sitting in the corner thinking about stuff.

"When I turn 60, I'm going to be less judgmental and more open minded," she mused. "I'm going to stop being a cowboy . . . no, I don't want to stop being a cowboy . . . I don't know . . . it's hard. Cowboys can be deep."

And with that she strode out to a taxi and rode off into the Roman sunset, continuing the scientific quest to know what else is out there to tame.

THE CELLO OR THE TRUMPET?

Are the quarks that Franklin helped discover the final frontier, or might they one day divide into even smaller pieces, just as the atom broke into electrons, protons, and neutrons, which in turn broke into quarks?

Many physicists feel that current experimental evidence, combined with the mathematical theory that underpins these experiments, has given us the answer to the question of the true indivisible units of all matter. Just as the periodic table of 118 chemical elements could be reduced to different ways of assembling the three basic building blocks of the electron, proton, and neutron, the hundreds of new particles found in the cosmic-ray collisions could be reduced to a simple collection of ingredients. The wild menagerie of particles has been tamed. But how sure am I that the gates won't open again to new customers? The truth is that physicists don't know.

If you look at the symmetrical model underlying these particles, then the triangle corresponding to the quarks is the last indivisible layer. The mathematics of symmetry suggests that we've reached the end of the road. That triangle corresponding to the quarks is the indivisible layer that builds all the other layers. So the mathematics of symmetry tells me I've hit the indivisible. Maybe we are falling into the same trap that Gell-Mann did when he first dismissed quarks because they had fractional charges. And yet another feature of quarks and electrons provides some justification for the belief that they might not come apart: they don't seem to occupy any space but behave as if they are concentrated at a single point.

In mathematics, geometry is made up of three-dimensional solids, two-dimensional planes, one-dimensional lines, and zero-dimensional points. The strange thing is that these were meant to be abstract concepts that didn't have a physical reality in our three-dimensional universe. After all, what is a line? If you draw a line on a piece of paper and look at it under a microscope, you'll see that the line actually has width. It isn't really a line. In fact, it even has height, because the atoms that are sitting on the page are piling up to create a little ridge of graphite across the page.

Similarly, a point in space might be identified by its GPS coordinates, but you wouldn't expect to find an object located solely at this point and nowhere else. You could never see it. Its dimensions would be zero. And yet an electron behaves in many ways as if it is concentrated in a single point in space, as do the quarks inside the proton and neutron. The manner in which electrons scatter off one another and off the quarks makes sense only in the context of a model in which these particles have no volume. Give them volume and the scattering would look different. If they truly are point particles, you wouldn't expect them to come apart.

But what about the fact that electrons have mass? What is the density of an electron? It should be mass divided by volume. But what if the volume is zero? Divide by zero and the answer is infinite. Infinite? So is every electron actually creating a tiny black hole? We are firmly in the territory of the quantum world, because where a particle is located turns out not to be as easy a question to answer as one might expect, as we shall discover in the next Edge.

Have the discrete notes of my trumpet won out over the continuous glissando of the cello? It is very difficult to know. Atoms were regarded as indivisible because of the indivisible nature of the whole numbers that showed how they combine. And yet they eventually fell apart into the tiny pieces that make up our conception of the universe. Why shouldn't I expect history to repeat itself with more surprises as I dig deeper and deeper? Why should there be a beginning, a first layer that made it all? It's a classic problem of infinite regression that we shall meet over and over again. As the little old lady once told a scientist who mocked her theory that the universe was supported on the back of a turtle, "You're very clever, young man, very clever, but it's turtles all the way down!"

Even if electrons and quarks are particles concentrated at a single point in space, there is no reason why a point can't actually be pulled apart to be made of two points. Or perhaps there are hidden dimensions we have yet to interact with. This is the suggestion of string theory. Point particles in string theory are actually believed to be one-dimensional strings vibrating at resonant frequencies, with different frequencies giving rise to different particles. Perhaps my cello does win out over the trumpet and the fundamental particles are really just vibrating strings.

Will we ever find ourselves at the point at which there are no new layers of reality to reveal? The current theory of the very small—quantum physics—proposes that there are limits to knowledge as I try to divide my dice into smaller and smaller units, that at some point I will run up against a barrier beyond which I cannot pass, as my next Edge reveals.

THIRD EDGE

QUANTUM PHYSICS

5

It is absolutely necessary, for progress in science, to have uncertainty as a fundamental part of your inner nature.

—Richard Feynman

I T'S EXTRAORDINARY what you can buy over the Internet. Today a small pot of radioactive uranium-238 arrived in the post. "Useful for performing nuclear experiments," the advert assured me. I rather enjoyed the comments from other people who purchased a pot: "So glad I don't have to buy this from Libyans in parking lots at the mall anymore." One purchaser wasn't so happy: "I purchased this product 4.47 billion years ago and when I opened it today, it was half empty."

Uranium is a naturally occurring element, and I'm assured it is safe to have on my desk next to me as I write. The documentation just warns me that I shouldn't grind it up and ingest it. The packet claims that it emits radiation at a rate of 766 counts per minute. It's kicking out a range of radiation: alpha particles, beta particles, and gamma rays. But what the specifications cannot guarantee is when exactly the uranium is going to spit out its next particle.

In fact, current quantum physics asserts that this is something I can never know. There is no known mechanism so far developed that will predict precisely when radioactive uranium will emit radiation. The Newtonian physics that I explored in the First Edge implied that theoretically everything in the universe is controlled by and follows a deterministic set of mathematical equations. But at the beginning of the twentieth century a group of young physicists—Heisenberg,

Schrödinger, Bohr, Einstein, and others—sparked a revolution, giving us a new perspective on what we can really know about the universe. Determinism was out. Randomness, it appears, rules the roost.

To understand this new frontier, I will have to try to master one of the most difficult and counterintuitive theories ever recorded in the annals of science: quantum physics. Listen to those who have spent their lives immersed in this world and you will have a sense of the challenge we face. After making his groundbreaking discoveries in quantum physics, Werner Heisenberg recalled, "I repeated to myself again and again the question: Can nature possibly be so absurd as it seemed to us in these atomic experiments?" Einstein declared after one discovery, "If it is correct it signifies the end of science." Schrödinger was so shocked by the implications of what he'd cooked up that he admitted, "I do not like it and I am sorry I had anything to do with it." Nevertheless, quantum physics is now one of the most powerful and well-tested pieces of science on the books. Nothing has come close to pushing it off its pedestal as one of the great scientific achievements of the last century. So there is nothing to do but to dive headfirst into this uncertain world. Feynman has some good advice for me as I embark on my quest: "I am going to tell you what nature behaves like. If you will simply admit that maybe she does behave like this, you will find her a delightful, entrancing thing. Do not keep saying to yourself, if you can possibly avoid it, 'But how can it be like that?' because you will get 'down the drain,' into a blind alley from which nobody has yet escaped. Nobody knows how it can be like that."

The revolution these scientists instigated is perfectly encapsulated in my attempts to understand what my pot of uranium is going to do next. Over a long period of time the rate of radioactive decay approaches a constant and on average is very predictable, just like the throw of my casino die. But according to the physics of the twentieth century, there is a fundamental difference between the die and the pot of uranium. At least with the die I have the impression that I could know the outcome, given enough data. However, there seems to be no way of determining when the uranium will emit an alpha particle. Complete information doesn't make any difference. According to the current model of quantum physics, it's completely and genuinely random. Quantum physics demolished Laplace's theory of a clockwork universe.

For someone on the search for certainty and knowledge, the revelations of quantum physics are extremely unsettling. There is nothing I can do to know when that pot on my desk is going to emit its next alpha particle? That's deeply shocking. Is there *really* no way I can know? There is much debate about whether this is truly something random, something we can never know, or whether there is a hidden mechanism we have yet to uncover that would explain the moment that radiation occurs.

This unknown is related to an even deeper layer of ignorance shrouding the universe of the very small. In order to apply the equations of motion Newton discovered to calculate the future evolution of the universe, it is necessary to know the location and momentum of every particle in the universe. Of course, practically speaking, this is impossible, but discoveries made in the twentieth century hint at a deeper problem. Even if I take just one electron, it is theoretically impossible to know precisely both its position and its momentum. Our current model of the very small has a built-in limitation on what we can know. This is called Heisenberg's uncertainty principle.

Whereas my First Edge revealed that the randomness meant to describe the roll of the die is just an expression of a lack of knowledge, the world of the very small seems to have randomness at its heart. I have learned to cope with the unknowability of the roll of the die, as deep in my heart I know it is still dancing to the regular beat of Newton's equations. But I'm not sure I can ever come to terms with the unknowability of my radioactive pot of uranium—something that apparently is dancing to no one's beat. Will it always remain unknowable, or does another theoretical revolution need to take place, akin to the radical new perspective that emerged at the beginning of the twentieth century?

WAVE OR PARTICLE?

The first inklings of this revolution came when scientists tried to understand the nature of light. Is it a wave or a particle? Newton's great opus on optics, published in 1704, painted a picture of light as a particle. If you considered light as a stream of particles, then the behavior that Newton described in his book appeared very natural. Take the way that light reflects. If I want to know where a ray of light will emerge when it hits a reflecting surface, then thinking of it like a billiard ball fired at the wall gives me a way to predict its path. The geometry of light, Newton believed, could be explained only by thinking of light as composed of particles.

Rivals soon emerged who believed that a wave was a much better model for describing the nature of light. Too many characteristics of light were hard to explain if it was a particle. An experiment that English physicist Thomas Young concocted in the early 1800s seemed to be the nail in the coffin for anyone believing in light as a particle.

If I shine a light at a screen with a single thin vertical slit cut into it, and place a photographic plate beyond the screen to record the light as it emerges, the pattern I observe on the photographic plate is a bright region directly in line with the slit and the light source that gradually tails off to darkness as I move away from this central line. So far, this is consistent with a particle view, in which small deflections can occur as the particles pass through the slit, causing some of the light to fall on either side of this bright region. (Even with the single slit, if the slit is small, there is some wavelike variation in the intensity of the light as you move away from the central bright region, which hints at waves at work.)

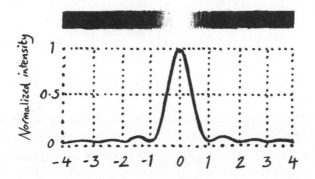

The intensity of light recorded on the photographic plate after the light has passed through a single narrow slit

The trouble for the particle theory of light was revealed when Young introduced a second vertical slit into the screen, parallel to the first slit. You would have expected to see two bright regions occurring in line with each slit corresponding to the particles of light passing through one slit or the other. But that's not what Young observed. Instead, he found a whole series of bands of light and dark lines across the photographic plate. Bizarrely, there are regions of the plate that are illuminated if only one slit is open and plunged into darkness if two slits are open. How, if light is a particle like a billiard ball, can giving it more options result in its suddenly being unable to reach certain regions of the plate? The experiment truly challenged Newton's model of light as a particle.

Only a model of light as a wave could possibly explain these bands of brightness and darkness. Think of a still lake full of water. If I throw two stones into the water at the same time, then the waves caused by the stones will interact in such a way that some parts of the waves combine to form a much larger wave and other parts of the waves cancel each other out. If I place a piece of wood in the water, I

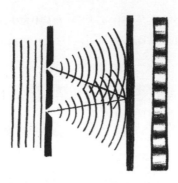

Light emitted from the left passes through the double-slit screen and strikes the photographic plate on the right. The light and dark bands depicted to the right of the plate represent the interference pattern detected.

will see this interaction as the combined waves strike the board. The wave hits the board with a series of peaks and troughs across the length of the board. The same principle explained what Young had observed. The light emerging from the two slits appears to produce two waves that interact in a similar way to the stones thrown into the water. In some regions the light waves combined to create bright bands, while in others they canceled each other out to produce dark bands. No particle version of light could get anywhere near explaining these patterns.

Supporters of the particle theory finally had to throw in the towel in the early 1860s, when it was discovered that the speed at which light travels matched exactly the speed predicted by James Clerk Maxwell's new theory of electromagnetic radiation based on waves. Maxwell's calculations revealed that light was, in fact, a form of electromagnetic radiation described by equations whose solutions were waves of differing frequencies corresponding to different sorts of electromagnetic radiation.

But there was a twist in store. Just as Young's experiment seemed to push scientists toward a wave model of light, the results of two new experiments at the end of the nineteenth century could be explained only if light came in packets. That is, if it is made up of discrete bits or what scientists call "quantized," which is where the word *quantum* comes from.

A CACOPHONY OF WAVES

The first inkling that light could not be wavelike arose from trying to understand the electromagnetic radiation being cooked up in the coal-fired furnaces that drove the Industrial Revolution. Heat is movement, but if you jiggle an electron up and down, then, because it has a negative charge, it is going to emit electromagnetic radiation. This is why hot bodies glow: the jiggling electrons emit radiation. Think of the electron a bit like a person holding the end of a skipping rope. As the person's hand goes up and down, the rope begins to oscillate like a wave. Each wave has a frequency that records how often the wave pulses up and down per second. It's this frequency that controls, for example, what color visible light will have. Red light has a low frequency; blue light a high frequency. The frequency also plays a part in how much energy the wave contains. The higher the frequency, the more energy the wave has. The other contributing factor to a wave's energy is its amplitude. This is a measure of how big the waveform is from trough to peak. If you think of the skipping rope, the more energy you put in, the higher the rope will vibrate. For many centuries, scientists used the dominant frequency of the radiation as the measure of its temperature. Red hot. White hot. The hotter a fire, the higher the frequency of the light emitted.

I got a chance to see one of these coal-fired furnaces at work when I visited Papplewick Pumping Station near Nottingham. Once a month they fire up the furnaces for one of their "steaming days." The furnace is housed in a beautifully ornate Victorian building. Apparently the cost of building the station was so far under budget that there were funds left over to decorate the pump house. It feels like a church dedicated not to God, but to the science of the industrial age.

The temperature inside the furnace at Papplewick was hitting something in the region of a thousand degrees Celsius. Scientists at the end of the nineteenth century were interested in what the spectrum of frequencies of light looked like inside the furnace for any fixed temperature. A closed furnace can reach a thermodynamic equilibrium where the heat jiggling the atoms causes radiation to be emitted that is then reabsorbed, so none of the electromagnetic radiation is lost.

When the furnace hits equilibrium, what frequencies of radiation do you find inside it? You can think of this as like a lot of my cello strings waiting to be vibrated. The total energy of the vibrating string is a function of the frequency of the vibration and the amplitude of the vibration of the string. Higher-frequency waves need more energy to get them going, but this can be compensated for by creating a wave with lower amplitude. Classically, a fixed amount of energy can theoretically get waves of any frequency vibrating, but the amplitude will be correspondingly smaller as the frequency increases.

A theoretical analysis of the spectrum seemed to indicate that waves of arbitrary frequency would occur in the furnace. When I looked inside the furnace at Papplewick I didn't get zapped by a load of high-frequency X-rays. But the wave-based theory of electromagnetism predicted that I should. Not only that, but if I added up all the contributions of the frequencies inside the furnace at thermal equilibrium, then an analysis based on light as a vibrating wave would lead to the absurd conclusion that the total energy contained in the oven is infinite. The furnace in Papplewick wouldn't have lasted long if that had been the case.

At any given temperature, there seems to be some cutoff frequency beyond which the waves fail to get going and vibrate. The classical picture is the following: If light is like a vibrating cello string, then the oven should generate waves of all frequencies, the number of waves increasing with the frequency. At low frequency the graph is correct, but as the frequency increases, I see the intensity of the radiation at higher frequencies tailing off until, beyond some point (which depends on the temperature), no waves of greater frequency than this number are observed.

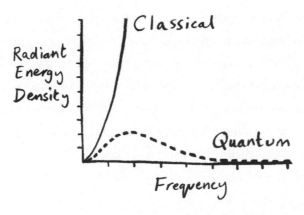

The frequencies inside a closed oven
predicted by classical and quantum models

In 1900, the German physicist Max Planck took the experimentally observed distribution of frequencies coming out of a furnace like that at Papplewick and came up with a clever idea to explain how to get the true curve rather than the nonsensical curve produced by a classical cello-string interpretation of light.

He posited that each frequency of electromagnetic radiation had a minimum energy that was required before it got going. You couldn't just continuously reduce the energy of a wave vibrating at a given frequency and expect it to sound. At some point, as the energy is reduced, the wave would just flatline. Planck's model asserted that there wasn't any continuous behavior. Each time the energy increased, it went up in jumps. The jumps in energy were tiny, so they were very hard to observe unless you were looking for them. But once Planck had this assumption in place, he found that the mathematical implications for the intensity of electromagnetic radiation at each frequency corresponded precisely with the observed radiation emerging from the oven.

So perhaps the universe was not the smooth, continuous place that scientists had believed it to be up to the end of the nineteenth century. Even the atomists—those who believed that matter was built of basic building blocks—had no idea whether this would apply to more intangible things like energy. If I draw the bow across my cello string and increase the pressure, although your ear hears a gradual and continuous increase in the volume, actually it is jumping up in steps. The size of the steps is very small. For any given frequency v, the energy, goes up in steps of $h \times v$, where h is called the Planck constant. This number, which controls the steps in energy measured in joule seconds, has thirty-three zeros after the decimal place before we get the first non-zero digit:

$$h = 6.626 \times 10^{-34} \text{ joule seconds.}$$

At this stage, Planck had no real physical explanation for the steps in energy. It was just what was needed mathematically to explain the electromagnetic radiation scientists had observed inside an oven like the one at Papplewick. It was Einstein's explanation of a second experiment that shifted scientists toward thinking of light as a particle rather than a wave. These particles each had an energy of $h \times v$.

THE PHOTOELECTRIC EFFECT

Metals are such good conductors of electricity because there are lots of free electrons that can move through them. This means that if I fire electromagnetic radiation at a piece of metal, I can actually kick these electrons off. The energy is transferred to the electron, which then has enough energy to escape the confines of the metal. This was the key to Thomson's discovery of the electron, described in the previous Edge.

If you think of electromagnetic radiation as a wave, I should be able to increase the energy of the wave until it knocks out the electron. The greater the energy in the wave, the more of a kick I will give the electron and the faster it will speed off. As I explained in the previous section, there are two ways to increase the energy of a wave like my vibrating cello string. One is to increase the frequency of the wave, to vibrate it faster. Do this, and, sure enough, the speed of the electrons that are ejected goes up proportionally. The other way is to increase the amplitude of the wave, to play it louder. The strange thing is that in this case, despite dialing up the intensity of the wave at a given frequency, the speed at which the electrons are emitted is not affected. What goes up is the number of electrons kicked off the metal.

If I decrease the frequency of the wave while increasing the amplitude, I can keep the total energy constant, but there is a point at which I can't seem to kick out any more electrons. There are frequencies below which, however loud I play my cello, the energy just doesn't kick out electrons. In contrast, with a high-frequency wave it doesn't matter how far down you dial the volume; even an extremely low-intensity wave has the power to knock out electrons. What's going on? How can I explain this strange behavior, which scientists call the photoelectric effect?

The answer is that we need to change our model. At the moment I've been thinking in terms of wave in, particle out. What if I try particle in, particle out? Perhaps the particle nature of the outgoing electron is actually the key to how I should view the incoming electromagnetic radiation.

This was Einstein's great paradigm shift, which he made in 1905, the year many call his *annus mirabilis*. It is also the year he came up with the special theory of relativity, which I will tackle in the Fifth

Edge when we explore the nature of time and the theory of Brownian motion, which provided the most convincing support for the thesis that matter is made from atoms.

Einstein suggested that we think of electromagnetic radiation, or light, not as a wave but as a machine gun firing off tiny billiard balls, just as Newton had suggested. The energy of each particle depends on the frequency of the radiation. With this new idea, we have a model that can perfectly describe what scientists were experiencing in the lab. In Planck's model of the energy inside the furnace, the energy did not go up continuously but went up in tiny steps equal to $h \times v$. What Einstein proposed is that each billiard ball of light will have energy corresponding to these tiny step changes. Electromagnetic radiation of frequency v should be thought of in Einstein's model as billiard balls each with energy $h \times v$. The jumps in energy that Planck introduced simply correspond to the addition of more billiard balls of light to the radiation. Einstein called these light balls quanta, but they were renamed in the mid-1920s, and we know them today as photons. How does this particle model of light explain the behavior of the electrons being kicked off a metal? Again, think of the interaction like a game of billiards. The photons of light come crashing onto the surface of the metal. If a photon hits an electron, the energy is transferred to the electron and the electron flies off. But the electron needs to receive a certain amount of energy before it can be kicked off.

The energy of each incoming photon depends only on the frequency of its radiation. If the frequency of the radiation is too low, the energy will be too small to kick out the electron. It doesn't matter if you turn up the intensity of the radiation; you're firing more billiard balls at the metal, but each individual billiard ball has the same energy. There is an increased chance of an electron being hit, but since each billiard ball is as impotent as any other, the electron is never going to be kicked off. In the wave model, the electron could sit there absorbing the incoming energy until it had accumulated enough to fly off. With the particle model I can kick the electron as many times as I want, but no single kick will be enough to knock out the electron. It's like gently prodding someone with your finger. Lots more gentle prods are not going to cause the person to fall over.

But if the frequency of the incoming radiation is above a certain value, the energy of each billiard ball will be enough to kick the electron off if it hits. It's like replacing hundreds of tiny prods with one huge shove: now the person falls over. Essentially, the ball has enough energy to transfer to the electron, and the resulting energy is sufficient to overcome the forces binding the electron to the metal. Increasing the intensity of the radiation means firing more balls at the metal, and this simply increases the number of electrons that will be emitted. Hence, rather than knocking out electrons with higher speeds, I just see more electrons being radiated off the metal.

The speed of the emitted electron in Einstein's model will be linearly proportional to the frequency. Interestingly, this relationship had not been observed or previously predicted, so it gave Einstein's model the perfect characteristic for any good scientific theory—not only could it explain what had been seen in the laboratory to date but it also predicted a new phenomenon that could be tested. This was important, as many scientists were intensely skeptical of Einstein's model. Maxwell's equations had been so persuasive that scientists were not going to change their minds without some convincing. How would Einstein explain the result of Young's experiment with the two slits and bands of light?

One of the skeptics was American physicist Robert Andrews Millikan. But his attempts to disprove Einstein's model of light as billiard balls of energy ended up confirming Einstein's prediction that the speed of outgoing electrons would be directly proportional to the frequency of the incoming radiation. Millikan had previously determined the charge on the electron as part of his research and would subsequently coin the term "cosmic rays," after he proved that the radiation being picked up by detectors on Earth was of an extraterrestrial nature. For all this work Millikan received the Nobel Prize in Physics in 1923, just two years after Einstein.

Einstein received his Nobel Prize in 1921 for his explanation of the photoelectric effect, not for his theory of relativity! Einstein's ideas gave the particle party a reason to retrieve the towel they'd thrown in a few decades earlier. Soon there would be a counter-revolution, which revealed that electrons have characteristics that make them look more like waves than discrete particles. It seemed as if light and electrons could behave both like particles and waves. Everyone was a winner in the new theory that was emerging!

Despite Einstein's paradigm shift, the experiments that were better explained by light as a wave were not invalidated. Weirdly, the circumstances of the experiments seemed to dictate which model of light you should use. The wave–particle duality was upon us.

Recall that Young's double-slit experiment had been the most devastating demonstration of why light is a wave, not a particle. The photoelectric effect had rather nicely used the particle nature of the electron to provide convincing support for thinking of electromagnetic radiation as a particle. But did the dialogue work the other way? What if I ask an electron to take part in Young's double-slit experiment? Firing electrons at a screen with two slits turned out to have shocking implications for our grasp on reality.

EXPERIMENTING WITH ELECTRONS

One of the most curious consequences of quantum physics is that a particle like an electron can seemingly be in more than one place at the same time until it is observed, at which point there seems to be a random choice made about where the particle is really located. Scientists currently believe that this randomness is genuine, not just caused by a lack of information. Repeat the experiment under the same conditions and you may get a different answer each time. This uncertainty about position is ultimately responsible for bits of my uranium suddenly finding themselves located outside rather than inside the pot on my desk.

The quintessential illustration of this repeats Young's double-slit experiment, but with electrons rather than light. A physicist colleague of mine in Oxford let me come and play in his lab so that I could see with my own eyes the bizarre game these electrons seem to be playing. I've read about it many times, but as Kant once said, "All our knowledge begins with the senses."

I felt compelled to warn my colleague that experiments and I don't mix well. No one would agree to be my lab partner at school because my experiments invariably went wrong. It was one of the reasons I was drawn to the more theoretical end of the scientific spectrum, where the mess of the physical universe can be kept at bay. But my colleague assured me this experiment was pretty robust.

To start, we set up a source emitting electrons at a rate that allowed me to record them arriving on the detector plate one at a time. I then placed the screen with two slits between the source and the detector plate. I first observed what happened when one slit was closed. The electrons that passed through the open slit hit the detector plate, and, after sending sufficiently many through, I began to see a pattern emerge.

The region directly in line with the source and the slit saw a high intensity of electrons arriving. As I moved to either side of this central line I still detected electrons arriving, but the number dropped off as I moved farther from the central line. The electrons sometimes seemed to be deflected as they passed through the slit, resulting in their paths being bent on either side of the central line. Nothing too strange up to this point. But then I opened the second slit.

If the electrons behaved like classical particles, I would have expected to see two regions of high intensity in line with the two slits, depending on whether the electron passed through the first or second slit. But this wasn't what I saw. Instead, I saw the interference pattern begin to build up, just as Young did when he shined light at the screen with two slits. What I observed was consistent with the analogy of a water wave passing through the slits and creating two waves that interfere with each other.

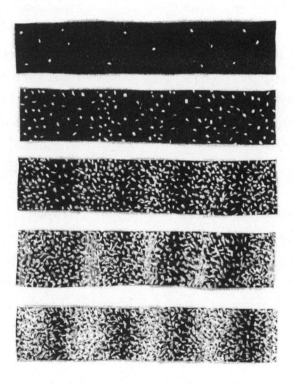

As more electrons are detected,
the wavelike pattern of interference emerges.

Remember that I'd set the experiment up so that only one electron was passing through the screen at a time. So there weren't many electrons interacting with each other in a wavelike manner. This was a single electron doing what a wave usually does. Even more inexplicably, there were regions on the detector plate totally devoid of any electrons arriving, despite the fact that with one slit open, electrons could reach this point. What was happening? I had opened another slit, providing several paths to a point on the detector plate, but the extra choice had resulted in no electrons arriving.

Kant proposed that all knowledge begins with the senses but "proceeds thence to understanding, and ends with reason, beyond which nothing higher can be discovered." So how did scientists distill reason from the strange behavior of these single electrons passing through the screen?

When it passed through one slit, how could the electron know whether the other slit was open? After all, the other slit was some distance away from the slit through which the electron was traveling. It's not that the electron splits into two and goes through both slits. The extraordinary thing is that I have to give up on the idea of the electron being located at any particular point until it is observed. To explain its strange behavior I need to describe the electron with a mathematical wave function that gives a range of possible locations. This was the revolutionary new viewpoint Austrian physicist Erwin Schrödinger proposed in 1926. The amplitude of the wave encodes the probability that when the electron is observed it will be found in a particular region of space.

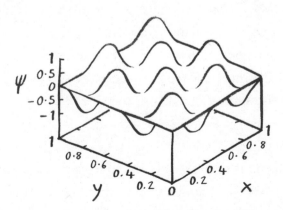

A quantum wave: the higher the wave, the more likely
it will be to find the electron at this point in space.

You might ask, a wave of what? What is vibrating? Is it a wave of information rather than physical stuff? Just as a crime wave isn't a wave of criminals but rather information about the likelihood of a crime happening in a particular area, the wave is simply a mathematical function, and a mathematical function is like a machine or computer—you input information, and it calculates away and spits out an answer. The wave function of the electron has as input any region of space, and the output is the probability that you will find the electron in that region. Amazingly, it should really be thought of as a piece of evolving mathematics. It is called a wave because the functions describing these probabilities have many of the features of classical wave functions. The peaks and troughs encode information about where the electron might be. The bigger the amplitude of the wave, the more likely you are to find the electron in that region of space.

When the wave encounters the screen with the double slits, it is affected by its interaction with the screen. The result is a new wave whose characteristics produce the strange interference pattern that I picked up on the detector plate. At this moment of detection, the electron must make up its mind where on the plate to land. The wave function provides the probability of where the electron might appear, but at this moment of detection the die is cast and probabilities become certainties. The wave is no more, and the electron looks once again like a particle hitting one point on the detector plate. Run the experiment again, and each time the electron might appear somewhere else. The more electrons I fire at the screen, the more I see the statistics encoded in the wave appearing. But in any individual case, physics claims I'll never know where on the plate I'll find the electron.

The curious thing is that I can return to the original experiment Young performed with light, and interpret it in light (if you'll excuse the expression) of Einstein's discovery. If I turn down the intensity of the light source in Young's experiment, I can reach a point at which the energy being emitted is so low that it corresponds to firing one photon of light at a time at the double-slit screen.

When this photon arrives at the photographic plate it shows up as a single point on the plate. So what's happened to Young's interference pattern? Here is the amazing thing—keep firing photons one at a time at the double-slit screen, and eventually the buildup of the pricks of light on the photographic plate reveals the interference pattern. Young wasn't witnessing a continuous waveform hitting the plate—it was an illusion. It's actually made up of billions of billions of pixels corresponding to each photon's arrival. To give you a sense of how many photons are hitting the plate, a 100-watt lightbulb emits roughly 10^{20} (or 100 billion billion) photons per second.

The wave quality of light is the same as that of the electron. The wave determines the probable location of the photon of light when it is detected. The wave character of light is not vibrating stuff like a wave of water but rather a wavelike function encoding information about where you'll find the photon of light once it is detected. Until it reaches the detector plate, like the electron, it is seemingly passing through both slits simultaneously, making its mind up about its location only once it is observed. If you are finding this hard to follow, don't despair. The scientists of the early twentieth century had an equally hard time trying to reset their model of how light or an electron really behaves.

It's this act of observation that is such a strange feature of quantum physics. Until I ask the detector to pick up where the electron is, the particle should be thought of as probabilistically distributed over space, with the probability described by a mathematical function that has wavelike characteristics. The effect of the two slits on this mathematical wave function alters it in such a way that the electron is forbidden from being located at some points on the detector plate. But when the particle is observed, the die is cast, probabilities disappear, and the particle must decide on a location.

I remember the Christmas when I first read about this crazy story of how things could appear to be in more than one place at the same time. Along with the toys and sweets that had been crammed into my Christmas stocking, Santa had also included a curious-sounding book, *Mr. Tompkins in Paperback* by a physicist called George Gamow. It tells the story of Mr. Tompkins's attempts to learn physics by attending a series of evening lectures given by an eminent professor. The trouble is, Mr. Tompkins always drifts off to sleep mid-lecture.

In his dream world, the microscopic quantum world of electrons is magnified up to the macroscopic world, and the quantum jungle Mr. Tompkins finds himself in is full of tigers and monkeys that are in many places at the same time. When a large pack of fuzzy-looking tigers attacks Mr. Tompkins, the professor who accompanies him in his dreams lets off a salvo of bullets. One finally hits its mark and the pack of tigers suddenly become a single "observed" tiger.

I remember being enchanted by this fantasy world and even more excited by the prospect that it wasn't as fantastical as it appeared. I was beginning to have doubts about the existence of Santa, given that he had to visit a billion children in the course of one night, but the book renewed my faith in the idea. Of course, Santa was tapping into quantum physics! Provided that no one actually observed him, he could be in multiple chimneys at the same time.

QUANTUM ANTHROPOLOGY

Now let us return to the double-slit experiment. If we try to sneak a look at which slit the electron "really" passed through by placing a detector at one of them, the interference pattern suddenly disappears. The simple act of looking changes the nature of the wave function charting the electron's trajectory. The pattern at the detector plate simply shows two regions of light lining up with the two slits. (The electron is either passing through one slit or the other and hits the screen behind the slit it passed through.)

Although it is a bit of a cheat, one way to think of this is to imagine an anthropologist observing a previously undiscovered Amazon tribe. To observe the tribe is to alter its behavior. It is impossible to observe without interacting in some fashion, and any interaction will affect the behavior of the tribe. This is even more evident in the case of an electron. To know which slit the electron went through there has to be some sort of interaction, and that interaction has consequences. If I look to see if the electron passes through one slit but don't detect it, I can infer that it has passed through the other slit. The detector hasn't interacted with the electron so much as with the wave function controlling the probability of where we'll find the electron. Before we looked, there was an even chance of the electron passing through either slit. By detecting no electron at one slit, we now know with one hundred percent certainty that it must have passed through the other slit.

Let's think of this in another way. Suppose I could make a bomb that would be activated by a single electron hitting a sensor. The trouble is, there is no guarantee the bomb works. The only way to know for certain is to engage in the rather useless act of firing an electron at the bomb. If it goes off, I know it works. If it doesn't, then it's a dud. But either way, after testing, I don't have a bomb.

The weird thing is, I can use the double-slit experiment to detect working bombs without setting them off. Remember that there were places on the screen that the electron can't hit if it is really going through both slits at the same time. We are going to use this region as the "bomb detector region." Place the bomb's sensor at the location of one of the slits, and if the bomb is a dud, the sensor won't detect the electron, and the electron can't hit the bomb detector region.

But what if it isn't a dud? Well, the sensor will detect the electron if it goes through the slit and sets the bomb off. Not much good. But because now I am detecting which slit the electron is going through, it will go through only one slit and thus has the chance to reach our "bomb detector region." So if I pick up an electron in the "bomb detector region," and the bomb wasn't detonated, it must mean the bomb is live.

Strangely, the act of observation can also stop the pot of uranium on my desk from decaying. By continually making lots of mini-observations, trying to catch it in the act of emitting radiation, I can freeze the uranium and stop it from decaying. It's the quantum version of the old adage that a watched pot never boils.

The code-cracking mathematician Alan Turing was the first to realize that continually observing an unstable particle could somehow freeze it and stop it from evolving. The phenomenon became known as the quantum Zeno effect, after the Greek philosopher who believed that because instantaneous snapshots of an arrow in flight reveal no movement, the arrow cannot in fact be moving at all.

Think of a particle that can be in two states, HERE and THERE. Unobserved, like the electron passing through both slits at the same time, we can consider it to be in a mixture of the two states, but observation forces it to decide which one. If it decides to be HERE, then after observation it begins to evolve into a mixed state again, but observe it quickly enough and it's still mostly HERE and will probably collapse into the HERE state again. So, by continually observing the particle, it never evolves sufficiently into the THERE state.

My children are obsessed with the science fiction TV series *Doctor Who*, just as I was as a kid. The aliens we find scariest are the Weeping Angels, stone figurines like those in our local cemetery that don't move, provided you don't take your eyes off them. But blink and they can move. The theory has it that the pot of uranium on my desk is a bit like a Weeping Angel. If I keep observing the uranium, which means a little more than keeping my eyes on the pot on my desk and involves something akin to surrounding it with a whole system of Geiger counters, I can freeze it in such a way that it stops emitting radiation.

Although Turing first suggested the idea as a theoretical construct, it turns out that it is not just mathematical fiction. Experiments in the last decade have demonstrated the real possibility of using observation to inhibit the progress of a quantum system.

Quantum physics seems to suggest that there are multiple futures for the electron in my double-slit experiment until I observe it, at which point a roll of the quantum dice determines which of these possible futures will occur. I suppose I can come to terms with the fact that I will never know the future until it becomes the present. After all, if I pick up my casino die with the intention of throwing it three times, there are $6 \times 6 \times 6 = 216$ possible future outcomes. My act of rolling picks out one of these 216 possible futures, just as observing the electron determines one of its many possible locations. But another twist of the double-slit experiment has the frightening implication that the past isn't uniquely determined either.

I seem, in fact, to be able to alter the past through my actions in the present.

It is possible to see which slit the electron went through long after it has passed through the screen. The observation device can be put up just before the electron is about to hit the detector plate. Let's call this observation device the "slit detector." Suppose I set up the double-slit experiment on a cosmic scale and I put my electron-emitting device on one side of the universe and the double-slit screen just in front of it, and my detector plate on the other side of the universe. It will take many years for the electron to travel through space until it finally hits the detector plate. When the electron passes through the screen with the double slit, it won't know whether I am going to observe it with my slit detector.

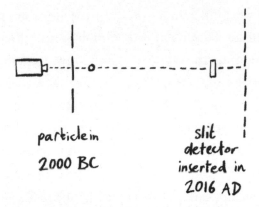

particle in
2000 BC

slit
detector
inserted in
2016 AD

**The decision to insert a slit detector in the path of the particle
in AD 2016 can alter what the particle did in 2000 BC.**

If, years later, I do use the slit detector to observe which way the electron went, it will mean that many years earlier the electron must have passed through one slit or the other. But if I don't use the "slit detector," then the electron must have passed through both slits. This is, of course, extremely weird. My actions at the beginning of the twenty-first century can change what happened thousands of years ago when the electron began its journey. It seems that just as there are multiple futures, there are also multiple pasts, and my acts of observation in the present can decide what happened in the past. As much as it challenges any hope of ever really knowing the future, quantum physics asks whether I can ever really know the past. It seems that the past is also in a superposition of possibilities that crystallize only once they are observed.

THE LIMITS OF REALITY

The interesting point for me—one that is often missed—is that up to the point of observation, quantum physics is totally deterministic. No one questions the nature of the wave equation that describes the electron as it passes through the slits. When he came up with his theory in 1926, Schrödinger formulated a differential equation that predicts the evolution of the wave function. Schrödinger's wave equation is as deterministic in some sense as Newton's equations for the laws of motion.

Uncertainty occurs only when I observe the particle and try to extract classical information. Suddenly I am left with an electron randomly located at some point in space. Over the long term, the randomness is described by information contained in the wave function, but scientists have not identified a mechanism that will let me determine where the electron will be located in each individual instance.

Schrödinger hated quantum physics' discontinuous changes in behavior. "If all this damned quantum jumping were really here to stay, then I should be sorry I ever got involved with quantum theory," he groused.

We should be careful not to over-egg the role of humans here. Worms, too, can presumably collapse the wave function. It is not just living creatures that are doing the measuring. There are particles on the other side of a potentially lifeless universe that are interacting with inanimate objects, causing the wave function to collapse. This interaction is as much an act of observation as my experimental investigations in the lab.

I am having real trouble getting my head around this idea of observation marking a divide between a deterministic electron described by a wave function and an electron that suddenly has a location determined purely by chance. The whole thing seems crazy. Nonetheless, there is no denying that it works as a computational tool. As the physicist David Mermin is reputed to have said to those who, like me, are unhappy with quantum logic, "Shut up and calculate."

Still, I can't help but feel that the objections are valid. The equipment I use to make a measurement is a physical system made up of particles obeying the laws of quantum physics. And I am too! I'm just a bunch of particles obeying quantum laws. Surely the observer, whether a photographic plate or a person, is part of the world of quantum physics, itself described by a wave function. The interaction of the wave function of the electron and the observer should still be described by a wave. After all, what constitutes an "observation" or "measurement"?

If the equipment and the observer and the particles are all described by a wave function, shouldn't everything be deterministic? Why are physicists happy to say that the act of observation collapses the wave function when frankly there is a mega-wave function at work describing all the particles at play—the electron, the equipment, and me. Where's the dividing line between the quantum world of probability and the classical world of certainties? This dualistic vision of a microscopic quantum world and a macroscopic classical world all seems a bit suspicious. Surely the whole shebang should be described by a wave equation.

It is all highly unsatisfactory, but the truth is that most physicists just take Mermin's advice and get on with it. My colleague Philip Candelas tells a story of how a promising young graduate student, for whom everyone had high hopes, suddenly dropped out of sight. When Candelas investigated what had happened, he discovered the reason. A family tragedy? Illness? Debts? None of these. "He tried to understand quantum mechanics."

I guess I have forgotten the advice Feynman gave to those who, like me, are quantum initiates: "Do not keep saying to yourself, if you can possibly avoid it, 'But how can it be like that?' because you will get 'down the drain,' into a blind alley from which nobody has yet escaped. Nobody knows how it can be like that."

That said, there are a number of ways in which people have tried to understand how it came to be like that. One is the hypothesis that at the point of observation, reality splits into a superposition of realities. In each reality, the photon or the electron is located in a different position so that the wave in some sense doesn't collapse but remains, describing the evolution of all these different realities. It's just that as conscious beings, we are stuck in one of the realities and unable to access the others.

This fascinating attempt to make sense of quantum physics is called the "many worlds" hypothesis, first proposed in 1957 by American physicist Hugh Everett. But are these many worlds really out there, existing simultaneously with our own? No one has yet come up with a way to probe these other worlds—or to prove whether they do, in fact, exist. The theory posits that there is just one wave function that describes the evolution of the universe in an entirely deterministic manner. It is back to Newton and Laplace, but with a new equation.

As a part of this wave function, we are denied access to other parts of it. It traps us inside, confined to one branch of reality. It may in fact be an in-built feature of our conscious experience that we can never experience these other worlds. But could I still use my math to analyze what is happening on the other branches? I can't see them, but can I at least describe them mathematically? Of course, just as the electron exists in many worlds, so do "I"—there are copies of me on the other branches witnessing the electron hitting other regions of the detector plate.

This model of reality is very intriguing, as it seems to affect directly what we understand by consciousness. I shall return to the question of consciousness in my Sixth Edge, but tantalizing questions are already dancing around in my head. Why am I aware of only one result of the electron hitting the plate? Is my conscious awareness of what is happening around me some version of the electron hitting a plate? Is the equipment in my skull unable to process multiple worlds? I look out of the window, and the photons emitted from the house opposite enter my eye and are detected on my retina. Why can't I sometimes look out and see house numbers 14 and 16 swapped over?

This attempt to reconcile what is going on in the physical world with what we are discovering at the subatomic level is challenging. Is it possible that the jump caused by the act of observation is not real but just something going on in the mind? We perceive a jump, but that isn't what is really happening. This unsatisfying explanation raises the question of what exactly we are trying to do when we give a scientific explanation of the world.

What is science? It is only by measuring and observing that we know anything. Mathematical equations can tell us what to expect, but it's just a story until we measure. So it's curious that our only way of "knowing" anything about the universe is to observe and make particles and light make up their mind about where and what they are doing. Before that, is it all just fantasy? I can't measure an entire wave function; I can only know it mathematically. Is a quantum wave function part of the universe that we can never really know?

Perhaps it's just greedy to believe that I can know more than I can measure. Hawking has certainly expressed such a view: "I don't demand that a theory correspond to reality because I don't know what it is. Reality is not a quality you can test with litmus paper. All I'm concerned with is that the theory should predict the results of measurements."

ONE INPUT, MULTIPLE OUTPUTS

My real problem with the current mainstream interpretation of quantum physics is that if you run the double-slit experiment twice, set up with exactly the same conditions, the outcome can be different. This goes against everything I believe in. I was drawn to mathematics by the certainty that a proof that there are infinitely many prime numbers means I'm not suddenly going to get finitely many prime numbers next time I check. I believed that science was made up of similar certainties, even if we as humans might not have access to them. I throw my die, and chaos theory tells me I may never be able to calculate the final outcome, but at least the mathematics says that if I start the throw in the same place it will end up with the same face pointing up. But quantum physics fundamentally questions whether this is the case.

Probability in the realm of dice is an expression of a lack of information. In quantum physics, even if I know everything, probability and chance remain.

Some might question whether it makes sense to talk about setting up the experiment and running it again with exactly the same conditions—that it is, in fact, impossible. Locally, you might get the conditions exactly the same, but you have to embed the experiment in the universe, and that has moved on. You can't rewind the wave function of the universe and rerun it. The universe is a one-time-only experiment that includes us as part of its wave function, and there's no going back.

But what if reality truly is random? Feynman, in his *Lectures on Physics,* states, "At the present time we must limit ourselves to computing probabilities. We say 'at the present time,' but we suspect very strongly that it is something that will be with us forever—that it is impossible to beat the puzzle—that this is the way nature really *is*."

It looks like the truly random thing sitting on my desk may not be the casino die I picked up in Vegas but that little pot of uranium I bought over the Internet.

6

How puzzling all these changes are! I'm never sure
what I'm going to be, from one minute to another.

—Lewis Carroll, *Alice's Adventures in
Wonderland*

I MUST ADMIT that I am having real trouble with the counterintuitive nature of the quantum world. Apparently, this is a good sign. The quantum physicist Niels Bohr once declared, "If quantum physics hasn't profoundly shocked you, you haven't understood it yet."

Richard Feynman went even further, declaring, "no one understands quantum physics." In a keynote address he gave in his sixties he admitted, "Might I say immediately that we have always had (secret, secret, close the doors!) a great deal of difficulty in understanding the worldview that quantum mechanics represents. I still get nervous with it."

The mathematician in me hankers after some mechanism that will tell me when my pot of uranium is going to spit out its next particle. Newton's equations held out the exciting possibility that if I know the momentum and position of a particle, then the equations of motion will tell me the complete behavior of that particle into the future. And if I repeat the same experiment with another particle located at the same point with the same momentum, it will repeat the path of the first particle.

The hope that we could ever really know the future was fundamentally crushed by the discoveries made by Heisenberg in 1927. He revealed that it doesn't actually make sense to say, "I know the momentum and the position of a particle at the same time." There seems to be an elastic relationship between knowing the location of a particle and knowing its momentum. If I measure the position of a particle with increasing precision, it seems to lead to its momentum having a whole range of different possible values. This is the content of Heisenberg's famous uncertainty principle. And as we shall discover, Heisenberg's uncertainty principle accounts for why the uranium sitting on my desk is randomly chucking out particles.

Heisenberg himself expressed well how important it is to be ready to reset your view of the world in light of new revelations: "Whenever we proceed from the known into the unknown we may hope to understand, but we may have to learn at the same time a new meaning of the word 'understanding.'" Quantum physics isn't about figuring out answers to old questions, but about challenging the questions we are allowed to ask.

Suppose I take one of the particles inside my uranium. If I know that the particle is at rest, it turns out I can't know where it is located. Indeed, there is a chance that when I look, I could find it anywhere across the universe. But if I try to pin down exactly where this particle is, I suddenly lose my handle on how it is moving. What appeared to be a particle at rest can suddenly find itself moving in any direction.

This seems totally crazy. If I throw my die through the air, I don't expect that watching it fall will suddenly cause it to fly off in a totally new direction. This intuition, as it turns out, applies only to things with a large mass. When the mass is as small as an electron, anything can happen. If I pin down the location of the electron to within the radius of an atom, its speed could change by as much as a thousand kilometers per second in any direction.

It's like trying to install a strange quantum carpet: every time I pin down one end of the carpet, the other end pops out.

To get a feel for this elastic relationship between position and momentum, let's return to my screen with the slits. I've been exploring the strange behavior of a particle fired at a screen with two slits in it. The tension between its position and momentum has actually already revealed itself in the behavior of the particle as it passes through a single slit. I get some diffusion of the particles as they pass through the single slit. But why should just one electron passing through the slit be deflected at all? Why doesn't it sail straight through? The trade-off between knowledge of position versus knowledge of momentum explains the diffusion I see.

If I can arrange for the electron to be fired from a distant source, which ensures that it passes through the slit, then I will know that there was no movement in the direction perpendicular to the slit. This means that as the particle enters the slit, I know the up–down momentum is zero.

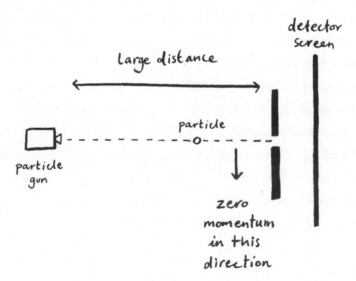

If I think of the electron as a point particle, then it either passes cleanly through the slit or it doesn't pass through it at all. If it passes through the slit, I have very precise knowledge of its position and should be able to predict precisely where it is going to hit the screen. The momentum was zero in the direction perpendicular to the slit before the electron entered the slit, so it should hit a region on the detector screen whose width is precisely the width of the slit. So why, as I fire more and more electrons through the slit, do I get the same diffusion pattern that I see for waves hitting the detector plate? Why aren't they all arriving in a region that is the same width as the slit?

Heisenberg's uncertainty principle asserts that any measurement that involves determining the precise position of the electron results in a new indeterminacy in the value of the momentum. So if the electron has passed through the slit, I can know the location of the electron within a margin of error given by the width of the slit. As the width of the slit is reduced, that margin of error decreases. But this causes the diffusion pattern to get wider and wider. Why? Because the value of the momentum is affected. While it was zero, once the electron emerges with its position narrowed down, the momentum becomes indeterminate. I've pulled the quantum carpet down in the position end and caused the momentum end to pop out.

This is a very strange situation. To top it off, the precise impact on the momentum is something I can't calculate in advance. I can measure it only at a later date. What I can know is a range of possible values in which I can expect to find the momentum once observed. Not only that, it now seems that if I repeat the experiment, I have only a probabilistic mechanism to determine what the momentum might be.

QUANTIFYING UNCERTAINTY

Heisenberg's uncertainty principle isn't some wishy-washy statement. It actually quantifies the loss of knowledge. Once I know the position of the electron with high precision, the momentum of the electron as it emerges can vary statistically around the average value of 0. I can't know what value I'll get when I measure the momentum, since that is still undetermined, but I know that statistically the possibilities for the momentum will be distributed on either side of the average value of 0. I can measure the spread of this distribution with something called the standard deviation, denoted as Δp, a statistical measure of the spread of possibilities. The greater the spread, the larger Δp, and the more uncertain I am of the value of the momentum.

Following Heisenberg's original paper detailing this strange inverse relationship between knowledge of position and knowledge of momentum, Earle Kennard and later Howard Robertson mathematically deduced the trade-off in knowledge. If the spread of possibilities of the position has a standard deviation of Δx, and the standard deviation for the spread of momentum is Δp, then these two values satisfy the following inequality,

$$\Delta x \, \Delta p \geq \frac{h}{4\pi},$$

where h is Planck's constant, which we came across when considering the energy of a photon of light. It is a mathematical consequence of quantum physics that the more knowledge you gain of the possible locations of a particle, the more this results in the possibility of the momentum being spread over a larger range of values. This is exactly what happens in the case of the electron passing through the single slit.

The order in which I do my measurements matters—the act of measuring position and momentum are described mathematically by two operations that give different answers if you do them in a different order. This idea can be illustrated using my casino die. Suppose I place my die on the table with the 1 on top, as illustrated on the next page. Now I am going to rotate it by a quarter turn around the vertical axis running through the top face and will follow that by a quarter turn through a horizontal axis running through one of the side faces. The top face now has a 5 showing. If I return the die to its original position and repeat the moves but in reverse order, horizontal axis spin followed by vertical axis spin, I get a different outcome. Now there is a 4 showing on the top face.

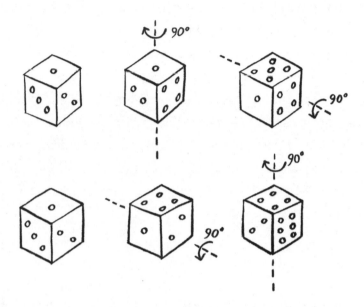

Any measurements that have this property—such that when you translate them into mathematical operations it matters what order you do them in—will give rise to an uncertainty principle. This is simply a mathematical consequence of a property called noncommutativity.

The math underlying quantum physics accounts for much of its counterintuitive nature. As I've buried myself in books and papers explaining quantum physics, I sometimes feel like I've entered a labyrinth. I thought I knew where I was before I started my journey. I've worked my way logically through the twists and turns of the labyrinth, relying on my math skills to lead me, as the walls are too high for me to intuit anything about the world beyond. The trouble is, once I make it out, where I've ended up doesn't look anything like where I started.

Quantum physics is a rabbit hole. Once you fall through it you have to reset your vision and formulate a new language to navigate this looking-glass world. For better or worse, that language is mathematics.

But can we trust the math? In this case, yes. The behavior predicted by Heisenberg's uncertainty principle has been confirmed. In a paper published in 1969, American physicist Clifford Shull described the

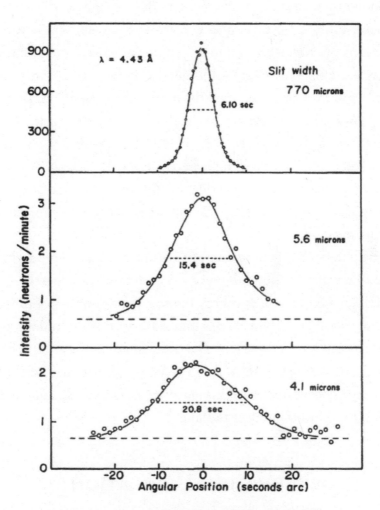

Clifford Shull's experiment confirmed that as the size of the slit is decreased, the statistical spread of the locations of the neutrons increases.

results of firing neutrons at slits of decreasing width. As the slit narrowed, the increased knowledge of the location of the neutrons resulted, as theory predicted, in a greater spread of possible values for the momentum. When the neutrons arrived at the detector plate, they were spread in a distribution whose standard deviation corresponded exactly to that predicted by the equation of Heisenberg's uncertainty principle.

An increase in the known must always be traded off against a corresponding increase in the unknown. But this indeterminacy can have some unexpected consequences. If I trap an electron in a very tiny box, its position is known to a high degree of accuracy. That results in the possible values of the momentum being spread over a huge range. As soon as I have measured the momentum, you might say that I now know both the position and the momentum. But it actually results in the position becoming indeterminate, so much so that I get an effect called quantum tunneling, whereby the particle that I thought was trapped inside a box can suddenly appear outside the box. This phenomenon is responsible for the radiation of the uranium that is sitting on my desk.

An alpha particle, consisting of two protons and two neutrons, can be found inside the nucleus of my uranium. In general, alpha particles don't have enough energy to escape the confines of the nucleus. But if we follow Heisenberg's uncertainty principle and think of the nucleus as a box, we will find that the position of these particles is not so clearly defined. In fact, there is a chance that they might be located outside the nucleus. This uncertainty about position is responsible for the possibility of radiation.

OBSERVATION IS CREATION

There have been many attempts to explain the uncertainty principle as a consequence of the act of observation. If you want to see where a particle is, you have to fire a photon at it, and this will kick the particle, giving it an unknown momentum. Beware of such explanations! They sound attractive, but they are misleading. In the description I gave above of the electron passing through a single slit, I did not need to fire any photons at the electron to alter its momentum. It was purely the act of the electron passing through the screen that gave me new knowledge of its location. There was no direct interaction with the particle that kicked it in one direction or the other. Apparently this misleading description of photons of light kicking the particle goes back to Heisenberg's original paper. He needed to include it in order to persuade his skeptical editors to publish the paper.

What does Heisenberg's uncertainty principle really tell us? Is it that we can never know the precise location and momentum of an electron? Or is it that such things don't exist? Or that observation is creation? His own formulation in this instance is that reality is created by our observations: "I believe that one can formulate the emergence of the classical 'path' of a particle as follows: the 'path' comes into being only because we observe it."

For some time I've found it very hard to believe that fundamental properties like position and momentum come into existence only once they are measured. The momentum of an electron might change after it passes through the tiny slit, but surely it must have some precise value even before I measure it? I can accept that I might not know what that value is until I determine the momentum, but there are lots of things I don't know before I measure them. Quantum physics, however, is trying to tell me that this belief in the existence of a precise value before measurement is a mistake. Can it really be my act of measurement that produces the reality of this particle?

It turns out I am in good company. Einstein (among others) tried to challenge the idea that things like momentum and position are indeterminate until they are observed. Surely, he argued, these things really do have explicit values as the particle flies through the vacuum. We might not know what they are, or have the machinery or knowledge to work them out, but they still exist. The argument is that we shouldn't confuse epistemology with ontology. I may not be able to know both momentum and position (epistemology), but that doesn't mean they don't exist (ontology).

I finally, reluctantly, had to give up on this intuitive belief when I read about Bell's theorem. In 1964, Northern Irish physicist John Bell definitively proved that it was impossible for certain properties of a particle to exist prior to a measurement being taken. Try to make them exist prior to measurement and everything ends in a contradiction. Bell showed that it was impossible to assign a value to account for every possible measurement without creating contradictory results. It's like trying to do a Sudoku puzzle with a mistake in it. However hard I try to assign values to all the squares, I will always get a row or column with the same number twice.

Given that Bell's theorem is as mathematically robust as they come, I've had to concede that the act of measuring truly seems to determine the properties of my particle. But I'm still deeply suspicious that the outcome is as random as current theory would have us believe.

I must admit to a sneaky feeling that quantum physics is a stopgap on the way to a more complete understanding of the behavior of fundamental particles. Surely there must be some mechanism deciding when my lump of radioactive uranium is going to spit out an alpha particle, or where the electron will hit after passing through the double slits.

Einstein certainly thought so, leading to his famous quote: "Quantum mechanics is very impressive. But an inner voice tells me that it is not yet the real thing. The theory produces a good deal but hardly brings us closer to the secret of the Old One. I am at all events convinced that He does not play dice."

Einstein believed that there must be something behind the veil that we have yet to penetrate. Even if we can't access it, he believed there must be some internal mechanism at work that determines the outcome. I am happy to concede that this mechanism may follow a model of randomness, just as the throw of my die does. But there must be something that determines the outcome. Perhaps the particles inside my pot of uranium have some little internal ticking clock, so that if the second hand is between zero and thirty when I measure it, the uranium will emit radiation; but if it's between thirty and sixty, then no radiation is detected.

The trouble is that a scenario first dreamed up by Einstein and his colleagues Boris Podolsky and Nathan Rosen revealed that if there is such a mechanism, then part of it could be on the other side of the universe. The scenario that Einstein, Podolsky, and Rosen cooked up involves the idea of quantum entanglement. It is possible to create two particles whose properties are entangled such that if I measure the properties of one particle, it will force the result to be mirrored by the other particle. It is a bit like having two casino dice, and whenever one lands on a 6, the other must too. It would be pretty difficult to rig up such an entangled pair of dice, but in quantum physics it is entirely possible to create such entangled and dependent particles. These entangled particles in turn reveal the very strange properties of any hidden mechanism that could determine how they will behave when measured.

To demonstrate the weird nonlocal nature of such a mechanism— if it exists—the experiment sends the two quantum dice in their entangled state off to opposite sides of the universe. If I then measure the first quantum die, it instantaneously determines which face is showing on the other at the opposite side of the universe. Some people, Einstein included, had a real problem with this "spooky" action at a distance. He felt that there might be a way for the roll of the dice to be preset before the particles head off to opposite ends of the universe.

But that was before Bell proved his theorem, which tells me that it is impossible to preset the properties of a quantum particle in advance of measurement. Remember—measurement is creation.

The real challenge is to understand how creation at one end of the universe can instantaneously create a new state for the second particle at the other end. Because if there is some internal mechanism at work determining the outcome of the second particle, that mechanism has just been altered by something that has happened on the other side of the universe. It cannot be localized.

Einstein already expressed his concerns at this "spooky action at a distance" with the double-slit experiment. How does the photographic plate know not to record an electron in one location if it is to be detected at a second location? There seems to be an instantaneous collapse of the wave function with no cascade effect from the point of observation across the result of the plate.

In this new case I have two particles, but because they are entangled in some sense, they are described by one wave function, which makes them not dissimilar to the particle being detected in the double-slit experiment. The two particles must be considered as one holistic unit. Bell's theorem means that the properties of the particles can't be preset before they travel to the ends of the universe, which in turn means that any mechanism determining their properties must span the entire universe.

So if there is a mechanism deciding when my radioactive uranium emits radiation (as my deterministic soul yearns for there to be), it will have to span the universe. It can't just be an internal mechanism in the heart of the blob of uranium on my desk.

Many scientists are not as keen as I am to eliminate the possibility that the behavior of my uranium is random, fueled by the hope that this small chink in the knowable will allow something many people cherish to enter the scientific picture: free will. Some have argued that if there is genuine randomness in quantum physics it is evidence of free will at work in the universe. Humans may not be totally free, but these microparticles seem able to do what they want . . . within reason.

Some religious thinkers contend that quantum physics allows room for an external agent to act in the world and influence its course. Provided that the results over the long term are in line with what we'd expect from randomness, there seems to be room for an agent to determine individual outcomes. This would take advantage of our present inability to explain how the macroscopic world of measurement interacts with the quantum world. So is the unknown of quantum physics home for a theistic God? If I wanted to get anywhere in my attempt to understand whether God could hide in the equations of quantum physics, I would need to talk to someone who was as much at home in a laboratory as in a cathedral. So I made a trip to Cambridge.

THE VEGETARIAN BUTCHER

John Polkinghorne learned his physics at the feet of Paul Dirac in Cambridge and then with Richard Feynman and Murray Gell-Mann at Caltech. You can't ask for better teachers than that. His research has, among other things, helped to confirm the existence of quarks. Polkinghorne is back at his alma mater, so I arranged to meet him at his home in Cambridge. Having done a five-year research stint at Cambridge, I always enjoy a chance to visit, even if my heart is with Oxford. Polkinghorne's decision to become an ordained priest, after a quarter of a century pushing the limits of quantum physics, makes him the perfect person to explore the theology in the quantum unknowable. To many it seemed like a dramatic career change. As he explained the decision, "I didn't leave science because I was disillusioned, but felt I'd done my bit for it after about 25 years. I was very much on the mathematical side, where you probably do your best work before you're 45."

Yikes. I hate it when people say that. I've always clung to the hope that it is a myth that mathematics is only for the under-forties. But then I guess, being on the wrong side of that divide, I would say that. So long as there are still unanswered questions to struggle with, that is what drives me on. And I've still got plenty of those unanswered questions on my desk. But I can certainly understand the desire to set yourself new challenges . . . like my current attempts to understand quantum physics. For Polkinghorne, the new challenge was getting ordained, and he often jokes about the seemingly contradictory natures of the two professions he has dedicated his life to. "People sometimes think that it is odd, or even disingenuous, for a person to be both a physicist and a priest. It induces in them the same sort of quizzical surprise that would greet the claim to be a vegetarian butcher." He himself thinks of the two roles as a harmonious combination. "The basic reason is simply that science and theology are both concerned with the search for truth."

I wondered whether there were any questions that he thought were beyond the reach of either discipline.

"There are two sorts of questions that science cannot answer," he offered. "Some of them arise out of science itself. The first is something we've learned from quantum physics, which is that although the world is orderly, it's also cloudy and fitful in its character and we don't have access to that clear, unquestionable post-Newtonian world that seems to be sitting there.

"But there are also questions that by their very nature don't lie within science's purview to answer. I think science has been tremendously successful and I have enormous respect for it, but it's achieved its success by limiting its ambition. Essentially, science is asking a single question about how things happen: What is the process of the world? And it deliberately brackets out by its nature questions of meaning and value and purpose."

That's not the first time I've come across this supposed dividing line. Science does the "how" and religion does the "why." It's an attractive sound bite, but I think it's a fundamentally flawed take on science.

Science tackles a lot of "why" questions. Why is my pot of uranium radiating alpha particles? Why do the planets orbit the sun on the same two-dimensional plane rather than at arbitrary angles to each other? Why do bees make their hives in hexagons? Why does the population of lemmings plummet every four years? Why is the sky blue? Why can't things travel faster than light?

Polkinghorne tried to tease out for me the difference he sees in the two approaches.

"My favorite homely example is that you come into my kitchen and you see the kettle boiling. If I put on my scientific hat, then I explain that it is boiling because the burning gas heats the water, and so on. But I can take off my scientific hat and say the kettle is boiling because I wanted a cup of tea and would you like to have one?"

I decided to take him up on his offer of tea. As it brewed, Polkinghorne continued.

"I don't have to choose between those two answers, and if I am to fully understand the phenomenon of the boiling kettle, I have to answer both questions: how it's happening and why it's happening."

I agree with Polkinghorne to some extent that science has limited its ambitions. Fermat's Last Theorem is frankly easier to grasp than the logic of my cat's behavior or the next move my son is going to make. But that doesn't mean that science can't hope ultimately to understand the complexities of a cat or the vagaries of human desire.

In my view, the science-versus-religion debate has fallen prey to our terrible desire to compartmentalize everything, the silo mentality that says, "This is science and this is theology and this is art and this is psychology." The exciting thing is that we have developed a multitude of discourses to navigate our environment. The evolution of everything in the universe might be reducible to the solutions of Schrödinger's wave equation—and that includes Polkinghorne's decision to boil his kettle—but while math is a great language for describing the behavior of my pot of uranium, it isn't the right language to explain the migration of a flock of birds, the thrill of listening to Mozart, or the immorality of torture.

Polkinghorne concurred that there were real dangers of a too-reductionist take on reality: "Sometimes when I'm having arguments with firmly reductionist friends who say that physics is everything, I say, first of all, 'What about mathematics?' and secondly, 'What about music?' Of course, music is just vibrations of the ear, but when you've said that you've said all that science can say about music, but you certainly haven't said all that can be said about music. It does seem to me very important that one doesn't just take a reductionist ax and chop everything down." I pushed him on his first example of a question science can't answer. Does he really believe that quantum physics means that I can't know when my pot of uranium is going to spit out its next particle? Is it really just chance?

"It's very unsatisfactory that there is this sort of lottery going on. A casino, in effect. Most quantum physicists who are busy doing the numbers have just got used to that, but I think it is unsatisfactory. The question is whether it is epistemic or ontological.

"Epistemological problems have an answer, but you don't happen to know it. But ontological things are situations where you could not know it. And that's the traditional interpretation of quantum theory— you cannot know.

"In the casino, we know it's essentially epistemic. There are tiny effects that influence things. My feeling is that if the problems of quantum theory are epistemic, then you need to have some notion of how that epistemological frustration arises, what stops you from finding an answer. I think it's sensible to try to push the issue ontologically as far as you can. We haven't got there yet."

Most quantum physicists believe that before you observe a particle, it is in a superposition of states described by the wave function, and that observation by macroscopic apparatus causes a jump in the behavior. The particle now has one state, and the wave function encodes the probability that you will find the particle in one state rather than another. There is no attempt to explain the jump. This is called the Copenhagen interpretation, after the home of its principal proponent, the Danish physicist Niels Bohr. Basically it's the "Shut Up and Calculate" school of quantum physics.

"Although I sign up to the Copenhagen interpretation of quantum theory, I don't think it's intellectually satisfying," Polkinghorne confessed. "At the end of the day all these things come down to someone saying 'and then it happens.' It's somehow produced by the intervention of macroscopic measuring apparatus. End of discussion. But that's just winning by definition. It is a problem. There are still puzzles."

Given Polkinghorne's belief that there is a God acting in the world, I wondered whether he thought the unknown of this collapsing wave function was a window for his God to act.

"I don't think that God is on hand to decide whether the nucleus in your uranium decays. There is some sort of mechanism . . . no, 'mechanism' isn't quite the right word . . . some sort of influence that sorts this thing out. One of the paradoxes of quantum theory is that here we are eighty years later and we still don't understand it."

I wondered why, as I discovered in the First Edge, Polkinghorne had chosen chaos theory, rather than his home territory of quantum physics, as the unknown through which his God might act.

"There was a period of about ten years when the science and theology communities were wrestling with these forms of agency. Of course they didn't solve the problem, because that would have been a very ambitious project. There were a lot of people, especially on the West Coast of America, who put their money on quantum theory explaining everything. That appeared to me just a little too slick. To counterbalance that, I lurched a bit too far in the other direction. I don't think chaos theory is the whole solution. It's really just the suggestion that the physical universe is orderly but looser in its order than Newton would have thought."

But he certainly isn't dismissive of the implications of quantum physics.

"The discovery of the intrinsic unpredictability in quantum theory shows us that the world certainly isn't mechanical and therefore we certainly aren't automata in some trivial and unbelievable sense."

It's intriguing to consider that an agent trying to dictate the course of the future using the unknown of quantum physics would have the opportunity to act only when measurement is made. Until the measurement causes a phase change, the equations of quantum physics are totally deterministic, rolling along in a linear, nonchaotic fashion, with no room for agency. This is one reason why religious physicists like Polkinghorne are not particularly enamored of the God of quantum physics.

As I drive back from Cambridge, I find myself lingering on this question of epistemology versus ontology, as it seems central to navigating what quantum physics tells me about the limits of human knowledge. Is it like my casino die? Although we can't know the precise starting point for the throw, we don't question whether the dice exist. But quantum physics questions whether I can talk about my pot of uranium as having a well-defined initial state. As Heisenberg put it, "The atoms or elementary particles themselves are not real; they form a world of potentialities or possibilities rather than one of things or facts."

Although Heisenberg's uncertainty principle seems to create an unknown or gap through which God can slip back in, it may actually fill another gap that is the spark for most people's belief in a creator. One of the big unknowns is this: Why is there something rather than nothing? My pot of uranium arrived in the post, sent via Amazon from Images Scientific Instruments in Staten Island. But if I keep reaching back, trying to find the ultimate origin of my uranium, I am eventually going to hit an unknown. The need for some explanation of this unknown is at the heart of many cultures' concepts of God. But what sort of answer is that?

I think most scientists who talk of God have in mind something that answers the seemingly unanswerable question of where all this stuff came from. Once the universe is up and running, they are happy to engage their scientific brains to understand how the stuff we have behaves. They are not looking for God to intervene in the world. This is what is often called deism rather than theism. This sort of God is very much one that can be equated with "things we cannot know."

Of course, if we try to describe what this answer actually looks like, we encounter the problem of infinite regress. If you think that something is responsible for creating the universe, then you quickly hit the question of who created that something. Of course, that "who" is part of the problem, because we have a terrible urge to personify this concept.

So this is why many talk about transcendent definitions, things that can't be articulated; to avoid the problem of infinite regress, they avoid even an attempt to articulate what the answer might look like. It's just something that is unknown and transcends our attempts to know it. This is the God the presenter of the Sunday morning program on BBC Northern Ireland whom I tangled with asked me about.

But if God is defined as something that can't be articulated, does it have any potency at all to act? If it can't interfere, or influence, if it can't be articulated and described, why do we need it? This is why mythmakers have had to mold their gods into forms that can be articulated, recognized, often personified. A God that is too transcendental loses its potency. For many early religions, this was precisely what happened to the idea of the High God or Sky God. As religious commentator Karen Armstrong writes in her book *The Case for God*, "He became *Deus otiosus*, a 'useless' or 'superfluous' deity, and gradually faded from the consciousness of his people."

As the theologian Herbert McCabe declared, "To assert the existence of God is to claim that there is an unanswered question about the universe." But he also warned that the fault of religion was always to make this God into a thing rather than a philosophical idea. The problem, he believed, is that religion far too often commits idolatry by trying to engage too personally with this concept of God.

The trouble is that an undefined, unknowable, transcendent concept is too abstract for many to engage with. It can't offer the sort of consolation that many seek. So perhaps it is inevitable that God's potency depends on becoming a little less transcendent, and more tangible.

ZERO EQUALS ONE MINUS ONE

The question of why there is something rather than nothing may not be as unanswerable as we think. As soon as you have a bit of empty space, quantum physics is going to start filling it with stuff. The version of Heisenberg's uncertainty principle that I have explored so far looks at the relationship between position and momentum. But there are other physical concepts that are similarly entangled.

For one thing, Heisenberg's uncertainty principle also connects the measurement of energy and time. If I look at what is happening in an apparently empty bit of space, then decreasing the time period in which I examine the space increases the uncertainty of the energy content—which means that empty space can never be truly empty. Over very short periods of time there is the chance of energy fluctuations. Since energy can change into mass, this results in particles spontaneously appearing from the vacuum. Most of the time, they annihilate each other and disappear back into the void, but sometimes things survive. And this gives us a mechanism for getting something out of nothing.

But where does this energy come from? Doesn't its sudden appearance contradict the concept of conservation of energy that physics holds dear? Some propose that the total energy content of the universe is actually zero, so no one is cheating the system. The key here is that gravity provides a negative energy content. So the universe can emerge from zero energy—from nothing—because what emerges is a combination of positive and negative energy. We are just seeing the equation $0 = 1 - 1$ at work.

It might seem a bit bizarre to call gravity negative energy, but think about putting a large mass like an asteroid next to the Earth. As the asteroid falls toward Earth it gains kinetic energy, but the gravitational pull is also going up because gravity increases the closer two masses are to each other. So to maintain a constant energy, this gravitational potential energy is negative and balances the increase in kinetic energy.

According to Heisenberg's uncertainty principle, it follows from the fact that space exists that you will get particles appearing from nothing. You don't have any need for a creator. Quantum fluctuations mean that we are seeing something appearing from nothing all the time. As we shall see in the Fifth Edge, this is how Hawking explained why black holes radiate particles. Nothing becomes a particle and an antiparticle—one gets trapped in the black hole and the other radiates away. So quantum physics already provides a partial answer to the something-from-nothing question.

However, you do at least need a stage on which to play this quantum game. Some equate empty space with nothing. But that is a mistake. Three-dimensional empty space, a vacuum, is still something. It is an arena in which geometry, mathematics, and physics can play out.

After all, the fact that you have a three-dimensional rather than a four-dimensional empty space already hints at the evidence of something. Nothing does not have a dimension. The ultimate challenge, therefore, is to explain how quantum fluctuations might produce space and time out of genuine nothingness.

FOURTH EDGE

THE UNIVERSE

7

I venture to suggest this solution to the ancient problem: the Library is unlimited and cyclical. If an eternal traveler were to cross it in any direction, after centuries he would see the same volumes were repeated in the same disorder. My solitude is gladdened by this elegant hope.

—Jorge Luis Borges, *The Library of Babel*

I HAVE ALWAYS BEEN INTRIGUED by the question of whether infinity physically exists. My attempts to create infinity by cutting the dice up into infinitely small pieces ran aground when I hit indivisible quarks. It even seems that I can't divide space infinitely, since space may be quantized. So my quest to know whether infinity exists will turn in a different direction.

What happens if I keep going in a straight line? Will I go on forever? It is a question that I think everyone who looks up into space must contemplate at some point in his or her life. If I throw my die into the vacuum of space, will it one day return to this starting point, or perhaps hit a cosmic wall and bounce back, or will it just tumble forever? Whether the universe goes on forever is a surprisingly subtle question that touches on the fact that space itself is not static.

To help me on my journey I've downloaded the universe—or what I can see of it at least. I've glued together a star globe from the European space agency website. It's not exactly a globe. It's made from two sheets of A4 paper that I've cut out and glued together to form one of my favorite mathematical shapes—an icosahedron made from 20 equilateral triangles. Like my casino die, it is an example of one of the five Platonic shapes that make good dice.

As we look out into space on a clear night, it seems as though the stars are all painted onto a huge black celestial globe that encases the universe. This was certainly the model of the universe for many ancient cultures. They believed that the Earth sat at the center of this globe and that the globe would spin on an axis that passes through the Polaris star, the one star that seems to stay still in the night sky, while other stars rotate around it.

I've put my paper star globe on my desk with the Polaris star sitting at the top. Around the middle are the signs of the zodiac that mark the passing of the year, including, of course, my own constellation of Virgo. The sun appears to move through each of these constellations, taking a full year to return to its starting point. On the bottom of the shape are the stars that you can see from the southern hemisphere, the brightest of which is Alpha Centauri. This is actually made up of three stars, including Proxima Centauri, which is thought to be the nearest to our own star, the sun.

People have been making versions of my paper globe for millennia. Cicero writes of ancient Greek astronomers making models of the universe with stars marked on them. Sadly, none of the Greek models have survived, so I popped into one of my favorite museums in Oxford, the Museum of the History of Science, to see some others that have. There was a beautiful globe about half a meter high dating from the early sixteenth century. Made in Germany, it brings the constellations to life with graphic images of birds, fish, animals, and men printed onto paper sections that have been glued onto the sphere.

Although my modern paper cutout doesn't match the beauty of this sixteenth-century globe, its design harks back to Plato, who believed that the celestial globe that encases our universe may not be a sphere but a dodecahedron—another one of the Platonic shapes that make good dice.

TRIANGULAR TELESCOPES

Every culture has looked up into the night sky and contemplated what is out there. Most immediately obvious are the sun and the moon. But how could ancient cultures discover anything about these bodies, given that they were confined to the surface of the planet? For me this is one of the striking things about mathematics—it allows us to deduce things about the universe from the comfort of our observatories. The fact that we can know anything about regions of space we may never visit is extraordinary.

Trigonometry, the mathematics of triangles and angles, was a tool developed not to torture schoolkids but rather to navigate the night sky. It is actually the earliest form of a telescope. Already in the third century BC, Aristarchus of Samos was able to calculate the size of the sun and the moon in terms of the radius of the Earth and to determine their relative distance from the Earth, just with the help of a mathematical triangle.

When the moon is exactly half full, the angle between the Earth, the moon, and the sun is essentially 90 degrees. So by measuring the angle Φ between the moon, Earth, and sun I can use the mathematics of trigonometry to calculate the distance from the Earth to the moon relative to that from the Earth to the sun. The ratio of those two distances is the cosine of the angle Φ, and that is something that I can determine with a purely mathematical analysis.

The inaccuracy of the measurement meant that Aristarchus was off by a factor of 20. He'd estimated the angle to be 87 degrees, while the true value is 89.853, almost a right angle. Small differences in angles of this size result in quite large differences in the relative size of the lengths of the sides of the triangle. It would take the invention of the telescope and some more clever mathematics to truly determine the size of the solar system.

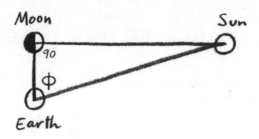

The right-angle triangle made by the Earth,
moon, and sun when the moon is half full

Yet even without telescopes, astronomers could see that the moon and sun weren't the only bodies processing through the night sky. Ancient cultures picked up several tiny pricks of light that behaved very differently from the plethora of other stars. They were wandering beacons of light—Mercury, Venus, Mars, Jupiter, and Saturn—luminous bodies that can't be marked on my cutout sphere because the next night they will be located at a different point. One explanation for the importance of the number seven to many cultures derives from the fact that with the sun and moon, the visible planets make up seven heavenly bodies.

Just as the planets move from day to day relative to the stars, it turns out that the stars are moving relative to each other. So the celestial sphere I have on my desk is just a snapshot of the night sky at a particular moment. Marked on my sphere is the easily identifiable constellation of the Plow or Big Dipper. But the stars that make up the Plow—Merak, Dubhe, Alkaid, Phecda, Megrez, Alioth, Mizar—are on the move: a hundred thousand years ago they would have been painted on my globe in a very different configuration, and a hundred thousand years hence they will look different again.

But for ancient astronomers the stars were fixed, bound to the celestial sphere that encompassed the universe. There was little discussion of what lay beyond them. Few medieval philosophers were prepared to contemplate the nature of the void. Nicolas Oresme believed that the extra cosmic space beyond the celestial sphere really existed and was infinite. In his writings he identified this immensity with God.

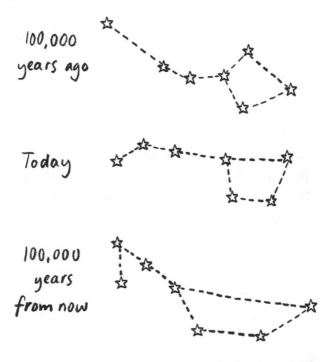

100,000 years ago

Today

100,000 years from now

The changing shape of the Big Dipper

Oresme was not frightened by the philosophical challenge of the infinite. Indeed, he proved that you can reach infinity by adding the fractions $1 + \frac{1}{2} + \frac{1}{3} + \frac{1}{4} + \ldots$, a counterintuitive mathematical result, given that the bits you are adding are getting smaller and smaller. This infinite sum is known as the harmonic series. When I pluck a string on my cello the sound is built from harmonics whose wavelengths are given by all these fractions. As I will explain later, the revelation that this harmonic series sums to infinity has interesting repercussions for how far we may ever be able to see into space.

It seems that it was only in the fifteenth century that astronomers began to contemplate the idea that the celestial sphere might be an

illusion, and that the universe might extend infinitely in all directions. Nicolaus Cusanus suggested that the universe was infinite and
any point could therefore be considered its center. This was picked up
by the Italian Dominican friar Giordano Bruno, who in 1584 wrote a
seminal work, *On the Infinite Universe and Worlds*: "The universe is
then one, infinite, immobile. . . . It is not capable of comprehension
and therefore is endless and limitless, and to that extent infinite and
indeterminable, and consequently immobile."

The logic that led Bruno to this conclusion is interesting. He argued that the universe is created by God, but God is unknowable.
Therefore the universe must be beyond our comprehension and must
be infinite, since a finite universe is, in theory, knowable. Now, I would
say that the converse is true—if the universe is endless, it means that
potentially it is beyond our comprehension. And if I am exploring the
concept of God as a way of articulating the things we cannot know,
then an infinite universe, if unknowable, could imply the existence of
this concept of transcendence. But is the universe infinite, and, if so,
is it as unknowable as it may first appear?

Bruno offers more than just his belief in God to justify his view of
an infinite universe. One of the most powerful challenges to a finite
universe contained in a celestial sphere is the question of what is behind the wall that contains the sphere. For centuries astronomers and
philosophers suggested that there was nothing—the void. But Bruno
was dissatisfied with this answer. He also believed that time was infinite both in the past and into the future. This somewhat controversially did away with the need for a moment of creation and a day of
final judgment. Bruno was not shy of controversy. His pronounce-
ments on the Bible eventually led to his falling foul of the Catholic
Church—not a good thing at the time. He was burned at the stake on
February 17, 1600.

How can we know if the universe is infinite? The opposite would
be much easier to prove. The surface of the Earth turned out to be fi-

nite and navigable, so perhaps we could simply navigate our universe and prove that it was finite. Although we don't yet have a ship in which we can sail to the edges of the universe, scientists in the seventeenth century did come up with a cunning way to explore space: the telescope.

HOW FAR CAN YOU SEE?

Galileo's generation discovered you could enhance how far you can see by placing carved glass lenses in a tube. For years Galileo himself was credited with the invention of the telescope, but that distinction really belonged to the Dutch spectacle-maker Hans Lippershey, who filed a patent for an instrument "for seeing things far away as if they were nearby." The Dutch instrument was able to magnify things by a factor of three.

Galileo first heard about the instrument on a trip to Venice. That same night he figured out the principle on which it worked, and he was soon constructing instruments that could achieve a magnification of thirty-three times. The name *telescope* was coined by a Greek poet who attended a banquet to honor Galileo in 1611; in Greek *tele* means "far" and *skopein* means "see." And Galileo's new contraption did indeed allow him and subsequent generations of astronomers to see farther than ever before. Galileo discovered moons orbiting Jupiter and sunspots that rotated, suggesting that the sun was spinning on its axis—phenomena that helped to confirm Copernicus's model of a sun-centered solar system.

In 1663, the Scottish mathematician James Gregory realized that the telescope could also be used to make new calculations of the distance of the sun from the Earth. Johannes Kepler had already observed the time it takes each planet to orbit the sun and deduced, using his laws of planetary motion, the relative distance of each planet from the sun. Kepler's third law stated that the square of the time a planet takes to complete one orbit of the sun is proportional to the cube of its distance from the sun. Venus, for example, completes an orbit in three fifths of the time it takes the Earth to go around the sun, which means that Venus's distance from the sun is about seven tenths (approximately $[3/5]^{2/3}$) of the distance of the Earth to the sun. (When we talk about distance from the sun we need to be careful since, as Kepler discovered, the planets aren't moving in perfect circles but in ellipses, so the distance actually varies. In general, I mean something like the average distance.)

But these are only relative distances. What Gregory and others realized is that they could use observations of the planet Venus crossing the sun, known as a transit, and a bit more trigonometry to calculate how far the Earth and Venus really are from the sun. If they made observations from two different locations on the Earth of the different points and times at which Venus crossed the sun, they would be able to work out the angle of the triangle formed by the two observers and Venus. And from there, by calculating the distance between the two observers on the Earth and using some trigonometry, they could deduce the distance to Venus.

The trick with trigonometry is to use triangles to change something you can't measure directly, such as the distance between the Earth and Venus, into something you can measure from the surface of the Earth, such as an angle or the distance between two points on Earth. The calculation was a complicated but clever application of abstract mathematical thought combined with practical astronomical observation.

The trouble is that these transits don't happen that often. Venus has crossed in front of the sun only ten times since 1400. Gregory had originally proposed using the transit of Mercury, since the next transit of Venus wouldn't be till 1761. Edmond Halley knew of his work and made observations of the transit of Mercury that occurred in

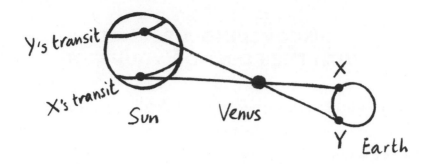

The transit of Venus from two different locations on Earth

1676, but it transpired that only one other observation was made—enough, theoretically, to calculate the distance, but given the errors that can creep in, you really want as many observations as possible.

It was the multiple observations of the transits of Venus in 1761 and 1769 that finally enabled accurate calculations to be made of the distance of the Earth from the sun. In one of the first coordinated global scientific experiments of its kind, it was calculated that the Earth was ninety-five million miles from the sun. Halley, alas, had died nineteen years earlier and missed the culmination of the project he'd tried to realize nearly ninety years before. Today, the calculations give the average distance of the Earth from the sun as 92,955,807.267 miles.

I continue to be astounded by the power of trigonometry to translate a measurement on Earth into the distance to a planet that no human has ever visited. More impressive mathematical feats were to follow. Telescopes and light aren't the only ways to detect what might be out there. Mathematics also turned out to offer a farseeing view of the universe, so much so that it succeeded in predicting the existence of a new planet before it was ever seen down the barrel of a telescope.

DISCOVERING A PLANET
WITH THE POINT OF YOUR PEN

There are two ways to find a new planet: luck and logic. The first planet to be detected since antiquity was discovered by luck. The German musician Friedrich Wilhelm Herschel had moved from Hanover to England in an attempt to further his musical career. He was also something of an amateur astronomer and he spent his nights staring at the sky with his impressive range of telescopes.

On March 13, 1781, Herschel spotted something unusual. What he at first thought was a star seemed to change size according to the magnification of his telescope. That was generally an indication of something close enough that it could be zoomed in on. The next test was: Did it move? Sure enough, four days later it had changed position relative to the stars. Given the number of comets that had been spotted, his first thought was that it must be a comet.

But after he alerted the Astronomer Royal and the object was tracked further, it became clear that its orbit was not parabolic, as you would expect of a comet, but nearly circular. The object was also too bright to be a comet and there was no visible tail. The astronomers concluded that it was in fact a new planet. Herschel wanted to name the planet after King George III, but classical mythology prevailed. Saturn was the father of Jupiter, and this new planet, orbiting farther out again, took the name of Saturn's father: Uranus.

Astronomers were ecstatic at the idea of a new planet and started mapping its trajectory, looking for moons, and calculating the length of its orbit around the sun. But Uranus was not as well behaved as they expected. Newton's theory of gravitation, so successful in predicting the path of the other planets, was distinctly faulty when it came to predicting where Uranus would be found. By 1788, the planet was $1/120$ of a degree out from where it should be. This led to the realization that the gravitational effect of Jupiter and Saturn needed to be factored in.

A new trajectory was published in 1791, but by 1800 Uranus was out of line again. By 1825, it was way ahead of its predicted position, but then it started to slow down, and by 1832 it had fallen behind the point where scientists had predicted it should be. Was there some mysterious substance out there causing resistance? Did Newton's laws of gravitation break down at such a great distance from the sun? Some suggested that just as Jupiter and Saturn were exerting a gravitational pull on Uranus, perhaps there was another planet out there pushing and pulling it around. But if another planet existed, where was it?

While Uranus was discovered by luck, the existence of this new planet would be pinpointed using the power of math and logic. Up until this point, astronomers would feed in the location of the planets and use mathematical equations to calculate their orbits. Now the process would be reversed. Uranus had been tracked for some decades. Where would a planet have to be to explain Uranus's strange trajectory?

The task was pretty daunting, but John Couch Adams in England and Urbain Leverrier in France decided to try their hand. Both succeeded in inverting the problem and identifying where they thought a new planet might be. Adams completed his calculations by September 1845 and approached astronomers in England to try to locate the new planet, but his lack of credentials and antisocial manner didn't endear him to the authorities. The Astronomer Royal was also distracted by a scandalous murder case involving one of his assistants. The net result was that Adams's prediction was ignored on the English side of the Channel. Leverrier took until June 1846 to complete his calculations and had similar problems persuading astronomical authorities in France to spend precious telescope time searching for a hypothetical planet. But he was more persistent, and wrote to the Berlin Observatory for assistance.

On September 23, 1846, Johann Gottfried Galle pointed the Fraunhofer telescope to the position in the night sky where Leverrier's calculations had predicted the new planet should be located. Sure enough, there was a point of light that wasn't on any of the star charts in the observatory. The next night it had moved . . . by exactly the distance Leverrier's calculations predicted.

The announcement of the discovery of a new planet was greeted with great excitement. But the astronomers who had failed to follow up on Adams's predictions were left looking rather foolish. Herschel, now a member of the Board of Visitors of the Royal Observatory, attempted to prove Adams had made the prediction first. The affair inevitably led to a bitter priority dispute. The naming of the planet became a political football, with the French wanting it named after Leverrier, and the English objecting that it broke with the tradition of naming planets after Roman gods. Eventually the international community settled on a name: Neptune.

As the astronomer François Arago wryly observed, Leverrier had discovered Neptune "with the point of his pen." Of course, Galle's observations at the Berlin Observatory had been required to confirm that theory did indeed match reality.

While we continue to be mostly stuck on the surface of the Earth, we've nevertheless managed to navigate to the edge of our solar system. How much farther would it be possible to explore? Although the telescope had allowed us to journey so far, it would also reveal the theoretical limits of how far we will ever be able to see—because, it turns out, light takes time to reach us.

COSMIC SPEED LIMIT

The telescope was instrumental in settling a dispute that had been raging since antiquity. Does light assert its presence instantaneously, or does it take time to travel from one point to another? Aristotle did not believe that light moved. He thought it was just present or not. The ancient Greeks thought that sight involved light traveling from the eye to the object. If that was so, argued the mathematician Heron of Alexandria, light must be instantaneous. Otherwise, when we open our eyes, how could we immediately see distant stars?

The Islamic scholar Alhazen was the first to suggest, in his *Book of Optics,* that light travels in the other direction, from the object to the eye. But even if light traveled the other way, many still believed that it would travel infinitely fast. Galileo wasn't sure. He thought that if light took time to travel from its source, then surely you could measure the time. His attempt to do so was too small in scale for any appreciable difference to be detected. Descartes revisited his experiment and increased the distance looking at solar and lunar eclipses. Both Galileo and Descartes were actually on to something—it's just that light travels so fast that the sun and moon were still too close to detect any discrepancy.

The moons orbiting Jupiter finally provided proof that it takes time for light to cross space. Galileo had proposed that the moons of Jupiter could be used as a clever means of solving the longitude problem. His method, which involved comparing the time of a lunar eclipse in different locations, didn't really catch on as a method for determining longitude at sea, which was the great challenge at the time, but it was used to determine longitude on land.

Then, in 1676, the Danish astronomer Ole Rømer used the moons of Jupiter to discover the finite speed of light. While working at the observatory in Paris, he recorded the times at which the moon Io disappeared into the shadow of Jupiter. The timing seemed to depend on where the Earth was on its orbit around the sun. If the Earth was on the other side of the sun from Jupiter, there seemed to be a delay. This, he realized, was because light took longer to travel to the Earth than if it was on the same side of the sun as Jupiter. Giovanni Cassini, the director of the observatory in Paris, announced the discovery on August 22, 1676, at the Royal Academy of Sciences in Paris. He explained that astronomers would need to change the tables predicting the times at which Io would disappear: "This appears to be due to light taking some time to reach us from the satellite; light seems to take about ten to eleven minutes to cross a distance equal to the half-diameter of the terrestrial orbit."

The current measurement is that it takes eight minutes and twenty seconds, so the seventeenth-century astronomers weren't far off. A whole series of experiments followed to determine the speed of light. Given that it is in the order of three hundred million meters per second, it is little wonder that many people thought it was infinite. Now, with the telescope making it possible to measure huge distances in space, the speed of light would become an important limit on any attempt to look into the far reaches of our universe.

I sometimes marvel, as I look out at the night sky, at the fact that I am looking back in time. A snapshot of the sky depicts the sun as it was eight minutes and twenty seconds ago, the nearest star as it was four years ago, and the most distant galaxies as they looked billions of years ago. It's possible that in some distant galaxy, telescopes trained on Earth are witnessing the extinction of the dinosaurs some sixty-six million years ago.

The speed of light became one of the ways astronomers measure the vastness of space. When astronomers say something is one light year away, they mean that it takes light one year to travel from that point to us.

THE STARS IN OUR NEIGHBORHOOD

Looking at the celestial sphere on my desk, I find it laughable that the ancient Greeks thought the stars were painted on a huge celestial ball enclosing the universe with the void beyond. But they didn't have much to go on. To the naked eye, the stars all seem to be a long way away. The invention of the telescope brought them a little closer, close enough for modern astronomers to see that they were not all the same distance from the Earth. Although we are stuck on the surface of the Earth, the Earth is moving relative to the stars, which means we get different viewpoints on the cosmos.

If you stick your finger in front of you and look out of the window and move your head from side to side, then you'll notice the things that are near, such as your finger, move a greater distance relative to the things that are farther away. This effect is called parallax. Astronomers can do the same thing when looking at the stars. If they compare the location of the stars in the summer to their position in the winter, this can tell them which stars are nearer to the Earth than others.

Herschel was trying to detect this stellar parallax, as it is now known, when he stumbled upon the new planet Uranus. The difference in locations of a star is extremely small, and it takes telescopes of a sufficient accuracy to detect the shift. Successful measurements weren't made until the 1830s, when the German astronomer and mathematician Friedrich Bessel recorded the first accurate observations of stellar parallax. To apply the technique to nearby stars, astronomers need to assume that the far-off stars are essentially on a single celestial wall that encases the universe, rather like the Greek model. The point is that for the purposes of detecting parallax of nearby stars, the farthest stars look fixed and can be used as a background to pick up the apparent movement of nearby stars.

Bessel compared the locations of a star called 61 Cygni in summer and winter. He made a triangle out of the star and two points in the Earth's orbit, and calculated one of the angles in the triangle. Then, taking his knowledge of the distance of the Earth from the sun and using the mathematics of trigonometry, he came up with the first estimate of how far the nearest stars were from the Earth. Bessel's calculation suggested that 61 Cygni was 660,000 times farther away from the Earth than the sun is. His calculation was about ten percent off the mark, but it was close enough to give us our first feeling for the depth of space.

Further calculations picked up other stars that were even closer. In 1915, Scottish astronomer Robert Innes spotted the closest known star to our own. Proxima Centauri is too faint to be seen by the naked eye, which may be why it took so long to detect it, but parallax calculations put the star 268,326 times farther away from the Earth than the sun, or 4.24 light years away.

Stellar parallax started to peel some of the stars off my celestial sphere and bring them closer to Earth. Provided a star is not more than 400 light years from us, this method works. But most stars still seemed to be so distant that they might as well be stuck on my paper sphere. Analysis of the wavelengths of the light coming from these stars let us take our next great step toward the edge of the universe.

TWINKLE, TWINKLE, LITTLE STAR

The farther away a star is, the less bright it appears to be. But how do we know whether I am looking at a bright star that is far away, like 61 Cygni, which can be seen with the naked eye, or a duller star that is closer, like Proxima Centauri? Apparent brightness is a combination of actual brightness and distance from the Earth. So how can astronomers use brightness to judge distances? It turns out that, in many cases, the color of the light emitted by a star gives us enough information to measure its distance from the Earth.

By measuring the light coming from a star and analyzing its frequencies, scientists saw that certain characteristic frequencies were missing. The light at these frequencies was being absorbed by the particular atoms in the star. This was key to proving Auguste Comte was wrong when he'd famously said that we'd never know the chemical composition of the stars. The light frequency could also be used to calculate how bright a star was. When astronomers looked at nearby stars whose distances from us (and thus their luminosity) were known, they found a direct relationship between the different frequencies absorbed by the star and how brightly it was shining.

This discovery meant that you could use the missing frequencies of light to measure absolute luminosity. Now astronomers turned again to the stars whose distances were too far away to apply parallax. By measuring the missing frequencies and their apparent luminosity, they could work out how far the stars were from us. This gave us for the first time a much clearer idea of the true depth of space.

In the end it was a very special, pulsating star that turned out to provide the best way to measure distances across the universe. Cepheid stars twinkle, and in 1912 American astronomer Henrietta Leavitt discovered how to use these twinkling stars to navigate the universe. She was employed at the time not as an astronomer but as a "computer" at the Harvard College Observatory, extracting data from the photographic plates for thirty cents an hour. Women weren't allowed to operate the telescopes. She'd been assigned the task of analyzing stars that grew brighter and dimmer over a period of time. Curious to know whether there was a pattern to the pulse of these stars, Leavitt focused on a batch of stars that were located in the Small Magellanic Cloud and were therefore all believed to be at a similar distance from the Earth.

When she plotted luminosity against the period of pulsation, she discovered a very clear pattern. The time it takes a Cepheid star to pulsate is directly correlated with its luminosity: the longer the period of pulsation, the brighter the star is shining. So to know how bright a Cepheid star really is, all you need to measure is the period of pulsation, something that is much easier to do than measuring missing light frequencies. These stars were perfect for measuring distances.

If a Cepheid star is pulsating slowly but looks very dim, it must be a long way away. Conversely, an apparently bright, rapidly pulsating Cepheid must look bright because it is close. With these new rulers to hand, the universe began to take shape. More stars took their place in the emerging map of the Milky Way. And our own star, the sun, we discovered, was tucked away in the corner of a huge spiraling mass of stars.

Was this the full extent of the universe? Had we finally seen and measured everything out there? Scientists had their doubts. There were some spots of light that didn't appear to be single stars. Were these clouds part of our Milky Way galaxy, or did they make up another galaxy like ours but completely separate? The first such region to come under investigation was a small cloud identified by the Persian astronomer Abd al-Rahman al-Sufi in the tenth century. It is bright enough to be detected by the naked eye and became known as the Andromeda nebula. In 1750, English astronomer Thomas Wright first suggested that this and other clouds might actually be galaxies in their own right. After reading about Wright's ideas, Immanuel Kant romantically referred to them as "island universes."

The debate about the status of these clouds raged for years, culminating in a face-off between rival groups in an event now called "The Great Debate," staged at the Smithsonian Museum of Natural History in 1920. At stake was the question of the size and extent of the universe. Astronomer Harlow Shapley presented the case for why such clouds would have to be part of our local galaxy to be shining so bright. Heber Curtis countered that the number of novae—cataclysmic nuclear explosions that stars can undergo—that had been counted in this cloud exceeded all those that had been recorded in the whole of our local galaxy. How could such a novae-rich region exist as part of our galaxy?

Andromeda was finally ripped out of our local galaxy and proved to be another galaxy in its own right by the observations of American astronomer Edwin Hubble, who in 1925 used the Hooker telescope at the top of Mount Wilson in California, the largest telescope of its kind at the time, to analyze how far away Andromeda was.

Hubble spotted one star in particular that he could use to calculate the distance. At the heart of the Andromeda cloud was one of the Cepheid stars that Leavitt had investigated. The star was pulsating, growing dimmer and brighter over a period of 31 days. Leavitt's analysis suggested that it was burning very brightly, and yet when viewed through a telescope it looked very dim. Combining the period of pulsation with a measurement of the apparent luminosity of the star revealed that it had to be 2.5 million light years away from our sun. The stars in the Milky Way had been calculated to be a maximum of 100,000 light years apart from each other. Leavitt's insight, combined with Hubble's calculations, dramatically changed our view of the universe. It turned out to be substantially bigger than anyone had previously imagined.

Leavitt's use of Cepheid stars as a way of navigating space transformed our picture of the universe—so much so that the Swedish mathematician Gösta Mittag-Leffler wanted to nominate her for the Nobel Prize in 1924. He was devastated to discover that Leavitt had died four years earlier from cancer.

The discovery of the existence of far-off galaxies gave us a new understanding of the true nature of space. But how far did the universe extend beyond these distant galaxies? The first explorers on Earth leaving their villages must have had the impression that the Earth was vast, perhaps extending forever. But as people journeyed more and more, there came a realization that the surface of the Earth was finite and navigable. So what about space? As we leave our galaxy, can we get a feel for how our cosmic village fits into a bigger picture of space?

A HUGE GAME OF ASTEROIDS

I can easily picture the Earth as finite but without an edge, a larger floating sphere. But how can space be finite? One of my favorite films that explores this puzzle is *The Truman Show*. It stars Jim Carrey as Truman Burbank, who doesn't realize that his whole universe is a scripted reality TV show housed in a huge dome. When doubts emerge as to the true nature of his world, he sets out in a boat across the water that surrounds his home town of Seahaven, only to discover that what he thought was an endless sky is actually painted onto the studio wall. Beyond the edge of his universe, he discovers cameras filming his every move.

I don't think we are living in our own *Truman Show*. I don't think that if I travel out into space I'll suddenly hit a studio wall or a celestial sphere like my model. And I think it's safe to assume most people would agree with me. After all, such a model only raises the question of what is beyond that boundary. Would we find a celestial film crew looking in on us? And what would be on the other side of that? This is why most people, when pressed, conclude that the only way to solve this conundrum is to believe in an infinite universe.

But mathematicians have a third alternative, which posits a universe that has no boundary but is nonetheless finite. In this universe you travel out into space until, rather than carrying on to infinity, eventually you find yourself heading back to your starting location, just like a terrestrial explorer rounding the Earth.

To get a feel for how this universe might work, it is useful to consider a small toy universe. The game of *Asteroids* created by Atari in 1979 serves as a perfect example of a two-dimensional universe that is finite yet without boundary. The universe consists of a computer screen, but when a spaceship heads toward the top of the screen, rather than bouncing back off the edge, like a 2D *Truman Show*, it seamlessly reappears at the bottom. As far as the astronauts inside the spaceship are concerned, they are traveling endlessly through space. The same applies if the spaceship heads to the left-hand side of the screen; it doesn't hit a wall but simply reappears on the right. The astronauts might begin to notice landmarks (or spacemarks) repeating themselves, although of course with an evolving universe it might be difficult to recognize things as you pass them for a second or third time.

This *Asteroids* universe actually has a recognizable shape. If I allow myself a third dimension within which to wrap up this universe, I can join the top and bottom of the screen to make a cylinder. Since the left- and right-hand sides of the screen are also connected, I can join the two ends of the cylinder to create a bagel or what mathematicians call a torus. The surface of this three-dimensional shape is the finite universe in the game of *Asteroids*.

If I take any finite three-dimensional shape, its two-dimensional surface will provide an alternate universe that is finite and unbounded. The surface of a sphere, for example, is another such two-dimensional universe. These two-dimensional universes are not just mathematical games but are key to navigating the Earth. Indeed, many cultures across the globe wondered whether the Earth went on forever or had an edge over which you would fall. Many ancient civilizations' model of the Earth consisted of a finite disk surrounded by water, a bit like Truman's world.

It wasn't until the fifth century BC that the Pythagoreans' idea of a spherical Earth started to take hold. The disappearance of ships over the horizon, the shadow created by the Earth on the moon during eclipses, the variation of the sun and stars in the sky as one travels south—all contributed to this shifting perspective. But for a long time it was nothing more than a hypothesis that went in and out of fashion. Copernicus, Galileo, and the astronomers of the Renaissance revived this idea, but it was only the 1522 circumnavigation of the world organized by Ferdinand Magellan (who was killed on the voyage) that finally proved beyond doubt that the Earth is round.

So what about the universe? Does it have a shape? We are sitting at a similar juncture to those ancient cultures contemplating the Earth and wondering whether it went on forever or had an edge or was somehow wrapped up.

How can I wrap up a three-dimensional universe so that it has finite volume but no edges? Such is the power of mathematics that it allows us to embed our three-dimensional universe in a higher-dimensional space and wrap it up like I did with the game of *Asteroids*. Although I can't physically picture the wrapping up, the language of mathematics gives me the equations to describe it and, more important, explore the properties of these finite three-dimensional universes.

So we may live in a three-dimensional version of *Asteroids*. Perhaps the universe is essentially a huge cube with six faces like my casino die. When a spaceship approaches one of these faces, instead of bouncing off the face it seamlessly exits the cube-shaped universe through one face and reappears at the opposite face. In the game of *Asteroids* two directions were joined up left–right and up–down. In my three-dimensional cube-shaped universe, the third direction is also fused. If I could put this cube in a four-dimensional universe I could wrap it up, joining the faces to make a four-dimensional bagel, or torus, whose three-dimensional surface is our universe.

But there are other possible shapes that our universe could take. A circle is a finite two-dimensional shape whose surface is a finite one-dimensional universe. A sphere is a finite three-dimensional shape whose surface is a finite two-dimensional universe. Mathematical equations can be used to construct a four-dimensional sphere whose surface is a finite three-dimensional universe—another model for what our universe could look like.

Although mathematics provides us with candidates for a finite yet unbounded universe, how can we ever know whether our universe is finite and what shape it might be? Do we have to wait for an astro-Magellan to circumnavigate the universe? Given the scale of the known universe, human exploration seems a rather hopeless way of proving whether the universe is finite. But there are explorers that have been navigating the universe for billions of years and can provide us with some insight into whether it is finite: photons of light.

COSMIC MAGELLANS

Light is a great explorer. We are constantly being showered with light that has spent billions of years traversing the universe. Does some of that light have a story to tell us that could give us an inkling of whether the universe is finite? I've already explored what would happen to a spaceship that plunged off into the depths of space; in these finite universes it would eventually come back to its starting position, just as Magellan's expedition did on its return to Seville in 1522.

The same could happen to light. Imagine a photon of light leaving our sun in its early years, some 4.5 billion years ago. Suppose we are living on the surface of a four-dimensional bagel, the cube-shaped universe where opposite faces are joined. What happens to the light as it approaches one of these faces? It passes seamlessly through, re-appearing at the opposite face, at which point it can continue its journey back toward its starting point. Provided nothing gets in the way, it could return and enter the telescope of an observer on Earth who detects this photon for the first time after its long journey. So what will our astronomer see? Well, nothing special. It will look like the light from a far distant star in its early years. It will be hard to detect that they actually are seeing what our sun looked like 4.5 billion years ago.

However, it does present us with a possible approach to proving that the universe is finite, because we can look in the opposite direction and see whether there might be a similar picture shining at us from the opposite face. Researchers in France, Poland, and the United States have been looking at the picture of light from the very early universe in the hope that parts of the map we've made may actually match up.

Much to their surprise and delight, they believed they had detected the first hint of some matching in the data. They began analyzing what shapes could possibly give rise to the patterns of wavelengths that they were observing. The results suggested that the best candidate for the shape of a universe that would produce these patterns is a dodecahedron. This is the dice-like shape made up of twelve pentagonal faces. Remarkably, Plato had suggested two millennia ago that the shape of the celestial sphere with the stars marked on it was not a sphere but a dodecahedron! In this modern interpretation, the suggestion was that like a joined-up cube, space would be fused at opposite faces of the dodecahedron. Interestingly, they had to give the pentagons a bit of a twist (by thirty-six degrees) before they could match them up. This is one theory, but the majority of astronomers are not convinced. It is difficult to tell whether the matchings are not the result of random coincidences.

There is another way that light could tell us about the geometry of our universe. Light tells us how the universe curves. Let's give our explorer setting out from his village a telescope and a featureless plain to venture across. At first the Earth seems flat, but after some time the curvature of the Earth becomes apparent: looking back, the explorer can no longer see his village—something is in the way. If the curvature continues across the whole surface, then it must join up to form a finite surface. Curvature like that of a sphere is called positive curvature. If a surface is flat, then it could be infinite, extending forever, but it could also be like the game of *Asteroids*, in which the universe on the screen is flat but finite. Flat surfaces are said to have zero curvature. There is another sort of curvature, which is like the saddle of a horse or a Pringle crisp. The curvature seems to dip down in one direction and up in the other direction. This is called negative curvature, in contrast to the positive curvature of the surface of a sphere. It gives rise to infinite surfaces, rather than finite surfaces like that of the sphere.

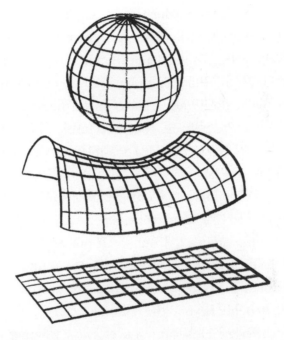

Two-dimensional surfaces with positive, negative, and zero curvature

Just as the two-dimensional surface of the Earth can be curved one way or another, it turns out that three-dimensional space also can have curvature. Measuring this curvature can give us some sense of how space might be wrapped up. If, like the surface of the Earth, the overall curvature of the universe is positive, the universe will curve in on itself into a finite shape. If it is negative it will be infinite. If it is flat, it could be infinite or finite like our cube-shaped universe fused at its opposite faces.

To pick up the overall curvature of space we can examine light as it traverses space. What do we see? Well, it almost looks flat, but it's hard to tell whether it is truly flat, or whether there is a small curvature. The margins seem to be so small that it is hard to say whether we'll ever be able to determine the curvature with sufficient accuracy to know which way space is bending.

There is another problem preventing us from really knowing the curvature of the universe. Much of our exploration of space depends on the assumption that where we sit in the universe isn't particularly special. It's called the Copernican principle. Once we thought we were at the center of it all. But Copernicus put an end to that. So we now believe that what the universe looks like around us is pretty much what it looks like everywhere. That is certainly what the evidence tells us. But it need not be the case. It's possible that the bit of the universe we see is rather special.

Suppose, for example, that our explorer on Earth had been living on a planet shaped like a hemisphere: perfectly flat on the bottom but suddenly curving off in a half sphere. If our explorer's village was on the flat bit, he'd think the whole planet was flat until he suddenly encountered a change in curvature. Perhaps the universe looks the same: flat in our patch, but beyond the region we can see, it does something completely different. How can we ever know whether the universe is as homogeneous as we think it is?

So the quest is still on to see if light is doing circles around a finite universe, like a cosmic Magellan expedition. If it is, this gives us the possibility that we may discover whether the universe is finite. Or perhaps it is bending in such a way that we can work out how the universe is wrapped up.

Magellan, of course, navigated a static planet. It turns out that our universe is a little more dynamic than we thought, as Hubble, the Magellan of space, discovered when he started analyzing the light coming from the stars in distant galaxies.

8

For in and out above, about, below
It is nothing but a Magic Shadow-Show
Play'd in a Box whose candle is the Sun
Round which we Phantom Figures come and go.

—Omar Khayyam, *The Rubaiyat*

I USED TO FANTASIZE about being able to look up at the night sky and point confidently at various stars and planets and declare: "That's Betelgeuse" or "You see that bright spot there, that's actually not a star but the planet Venus." But one of my problems is that I have an extremely bad memory. If something's random, like the stars scattered across the firmament, then without some logic to guide me, I find it very difficult to name much beyond the Plow. This is, of course, why we've created patterns like the Plow, or Orion the Hunter, to help us navigate these random spots of light.

But it turns out that I'm also not very physically suited to astronomy. My first attempt to take a deeper look into space was a trip to the telescope housed in Mill Hill in north London. But my wish to see to the edge of the universe was scuppered by the scourge of astronomers: clouds.

So next I tried to get above the clouds. This required more effort than jumping on the Northern Line to Mill Hill. I took a train to Switzerland, culminating in a beautiful alpine climb to the final stop at Jungfraujoch. From there a lift tunneled into the mountain took me to the mountain peak on which the Sphinx observatory sits, 3,571 meters above sea level.

Built in 1912, it looks like the villain's lair from a James Bond movie. As the sun set over the snow and glaciers, I prepared myself for a wonderful evening of stargazing. Except my body had other ideas. I'd already been feeling quite dizzy and even nauseous. I hadn't been able to eat much. As the first stars started to appear, I was hit by a crashing headache. Before long I was throwing up. When an elderly German couple said that I had all the symptoms of altitude sickness save one, it suddenly dawned on me that I'd never been this high before.

"And what's the last symptom?"

"Death."

It was then that I realized my dream of becoming an amateur astronomer wasn't really worth it. I got the first train back to a sensible altitude and the symptoms disappeared. I just had to face it: I'm a Thames Valley boy born in London whose body was built to stare up at the stars from the comfort of a telescope closer to sea level. But observations made by some of those great telescopes that sit high on the mountains across the world have revealed a remarkable fact. At some point in the future, there won't be so many stars for me to look at—they are disappearing over our cosmic horizon!

THE UNIVERSE THROUGH
RED-TINTED SPECTACLES

When an ambulance drives by with its siren blaring, your ears experience the squashing up of sound waves as the ambulance approaches, which causes the wavelength to shorten, resulting in the siren sounding higher-pitched than when the ambulance passes you. As it pulls away, the sound waves are stretched out, resulting in a longer wavelength and lower pitch. This phenomenon is known as the Doppler effect.

The same thing happens to light. As a star speeds away from us, its light shifts toward the longer red wavelength. If it moves toward us, the light shifts toward the shorter blue wavelength. Having already discovered that our galaxy was not special but just one of many, Edwin Hubble turned his attention in 1929 to analyzing the light from these galaxies to see how they were moving relative to our own. To his surprise, the light from distant stars in the galaxies he observed was all shifted toward the red. Nothing seemed to be coming toward us. It was as if the other galaxies were fleeing from us. Even more interestingly, the farther the stars were from our own star, the more the wavelength shifted. Hubble couldn't believe that the Earth was in such a special place in the universe. He soon realized there was a much better explanation: space is expanding in all directions simultaneously. Whatever your point of observation, it will appear that everything is moving away from you. The space between us and the stars is being stretched. The galaxies are being carried along by the expansion of space like leaves in the wind.

Although often credited to Hubble, an expanding universe was actually predicted by a Jesuit priest, Monsignor Georges Lemaître, two years earlier. Lemaître deduced that the universe must be expanding as a consequence of Einstein's equations for gravitation. When Einstein heard about this theory, he dismissed it with the damning declaration: "Your calculations might be correct but your physics is atrocious." Einstein was so confident that Lemaître was wrong that he ended up sticking something called the cosmological constant into his equations to try to force the universe to be static and thus scupper Lemaître's prediction.

Lemaître didn't help his cause by publishing his findings in an obscure Belgian journal. Once Hubble's observations supported the idea of an expanding universe, Einstein changed his tune. Hubble's and Lemaître's discoveries were the first indication that our universe isn't static. Instead, space is always on the move.

It is the stretching of space that makes the wavelength of the incoming light longer. The farther the light has traveled, the more its wavelength will gradually have been stretched. We call this stretching redshifting because the visible light with the longest wavelength is red. So to say that light has been redshifted simply means that the wavelength has been stretched. The bigger the redshift, the more space has been stretched on the way and therefore the farther the original source of the light from Earth.

To explain why this stretching changes wavelengths of light, blow up a balloon and mark three points on it. One represents Earth, the other two points are distant stars. Draw two waves of light of a fixed wavelength between the stars and Earth. This is what the light looks like when it leaves the stars. By the time the light from the nearer star reaches Earth the balloon has expanded. If I blow it up a bit, the wavelength gets longer. It takes longer for the distant star's light to reach

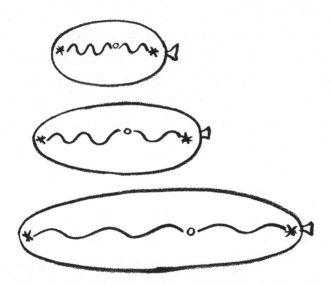

In the first balloon, the light leaves the stars with the same wavelength. In the second balloon, the universe has expanded and the light from the nearer star reaches Earth. The wavelength has gotten longer. In the third balloon, the universe has expanded further by the time the light arrives from the farther star. The wavelength has shifted even more toward the red.

Earth, so the universe has expanded a bit more. If I blow the balloon up even more, the wavelength gets longer again. So the greater the redshift, the farther the star is from Earth.

This was a new way of measuring distance to far-flung stars. It has helped astronomers to identify stars that must now be thirty billion light years away from us. It may seem something of a paradox that we can see something thirty billion light years away when, as we shall soon see, the universe is only 13.8 billion years old, but remember that we are seeing the star in the past, when it was close enough that the light could reach us. It is only by analyzing the expansion mathematically that we can infer that today the star is at a point from which it would take its light thirty billion years to reach us.

THE ANT AND THE BAND

What does it mean to live in an expanding universe? If the universe is expanding, then perhaps some stars will always remain out of sight, since they are moving farther and farther away from us. But scientists have discovered that if the universe is expanding at a constant rate, even if that rate is faster than the speed of light, if we wait long enough, we'll eventually see light from every star.

The best way to understand this is to consider an example that has a curiously counterintuitive outcome. Consider an ant (playing the role of a photon of light) sitting on the end of a rubber band (playing the role of space) that is stuck down at one end (playing the role of the Earth). The end where the ant starts plays the role of a far-distant galaxy.

Now pull the end of the rubber band at a constant speed. Suppose the rubber band starts with a length of one kilometer and every second it increases in length by one kilometer. The ant is moving at a much slower speed along the band, say, at one centimeter every second. At first glance, with the end flying off at a much faster rate, it appears as if the ant has no chance of reaching the other end, just as light from a galaxy far enough away seems to have no chance of reaching Earth if the space between them is expanding at a uniform rate.

$\frac{1}{100,000}$ of total

$\frac{1}{100,000} + \frac{1}{200,000}$ of total

$\frac{1}{100,000} + \frac{1}{200,000} + \frac{1}{300,000}$ of total

There are some subtleties here, so let me add a condition to help us get a grip on what is happening. The rubber band gets stretched only after each second, and it will be an instantaneous stretch. So after one second the ant has traveled one centimeter, which is 1/100,000 of the total distance. Then the band expands. The point to note is that although the distance that the ant needs to travel has increased, it has at least still covered 1/100,000 of the distance between the star and Earth, because the stretch has also helped push the ant on a little from its starting position.

Now the ant travels another centimeter. The rubber band is two kilometers long. So the ant has covered a further 1/200,000 of the total distance. The total proportion that the ant has now covered is 1/100,000 + 1/200,000 of this total distance. The band expands by another kilometer. The proportion hasn't changed with the stretch. The band is now three kilometers long. The ant travels another centimeter. This is only 1/300,000 of the total distance. Every time the rubber band stretches, the proportion that the ant's centimeter represents is going down. But hold on—here is the power of mathematics. After n seconds, the proportion of the band that the ant will have covered is

$$\frac{1}{100,000} + \frac{1}{200,000} + \frac{1}{300,000} + \ldots + \frac{1}{n \times 100,000} = \frac{1}{100,000} \left(1 + \frac{1}{2} + \frac{1}{3} + \ldots + \frac{1}{n}\right)$$

This is the harmonic series that the mathematician Oresme calculated many centuries ago and that I considered at the beginning of the last chapter. Oresme proved that this series will get arbitrarily large. So I can make n so large that the sum is bigger than 100,000. In that case, the proportion of the band that the ant has covered will be over one hundred percent. This means that the ant has arrived!

In this example I have the rubber band stretching at a uniform rate. This is pretty much how astronomers like Hubble thought space was behaving—perhaps even slowing down, given that gravity should have a decelerating effect. This implies that even if the universe is infinite, we should be able to see more and more of it if we just sit and wait—like a colony of ants crawling along the expanding rubber band of space.

Does this mean that in theory, if the universe is infinite, we should already be receiving light from all stars? Perhaps we can already see an infinite universe. The thing to remember, though, is that the farther away the star is, the further in the past we are looking at it. If we rewind our expanding universe far enough, we will find there were no stars at all.

REWINDING THE UNIVERSE

The discovery by Hubble and Lemaître of the expanding universe provided the first evidence for what scientists now call the Big Bang. If I reverse time, an expanding universe turns into a contracting universe. But keep contracting and the universe becomes so dense that the state of the universe changes rather dramatically. In fact, as Lemaître first realized, at some finite point in time this reversed expansion leads to a universe that is infinitely dense, a point he called the primeval atom or cosmic egg. This "singularity" is what scientists refer to as the Big Bang. Given that it is a point at which relativity and quantum physics must fuse into a coherent theory, there are still debates about how far we can rewind the universe before current models fail and new ideas are needed.

When I first heard about the Big Bang at school, I thought that if the universe started as a point then it had to be finite now. But with a bit of mathematics it is possible to show how an infinite universe can still start as a single point. This seems extraordinary. How can a point with no volume contain infinite space? To get an idea of how this works it is worth starting with an infinite space and working backward. Think of infinite space one second after the Big Bang. Take an arbitrary point as the center of this universe and consider all the points that are a distance R from this point. They all sit on the sphere of radius R.

Now I am going to rewind our universe backward to time zero. At time $t = \frac{1}{2}$, the points on the sphere of radius R will contract to points on a sphere of radius $\frac{1}{2}R$. At time $t = \frac{1}{4}$, they've contracted further to sit on a sphere of radius $\frac{1}{4}R$. As I keep halving the time in the direction of the Big Bang, this sphere is getting smaller and smaller, until at time $t = 0$ it becomes a point. But this is true of every sphere regardless of how big R is. So every point in my infinite space sits on some sphere of radius R and when rewound to $t = 0$ collapses to the point I chose. So in one second the mathematics gives me a way to suck infinite space into a single point with no volume. Shakespeare put it nicely when Hamlet declares, "I could be bounded in a nutshell and count myself a king of infinite space."

Of course, we run into problems with this model if space and time are quantized. As I saw when I tried to keep halving my die, there may come a point when I can't halve things. This is at the heart of the debate when quantum physics meets general relativity: trying to resolve what happens as we contract the universe to a point.

Many call this the beginning of our universe, and I shall return to the nature of what we mean by beginnings and time in the next Edge. But the Big Bang certainly has implications for just how far we may ultimately be able to see, because it means that the stars could not have existed for more than 13.8 billion years, which is the current estimate of how long ago the Big Bang occurred. It took some time after that before the first stars could form.

As you head out into space, you are heading back in time. Since there were no stars before 13.8 billion years ago, this means there is a sphere around us beyond which there will be nothing to see. The wonderful thing is that we are back to the model of the universe proposed by the ancient Greeks. There is a huge sphere with the Earth at its center, and photons from beyond that sphere have not had time to reach us yet. That sphere is getting larger as time passes, and the question of how much space is contained inside this expanding horizon will turn out to have an unexpected answer.

The farthest galaxy whose distance has been confirmed is one whose light has taken 13.1 billion years to reach us. That does not mean the galaxy is 13.1 billion light years away from us today, because the space between us has expanded over all those years. Calculations suggest that the galaxy is currently thirty billion light years away from Earth.

You might think we should be able to see light going back to the first moment after the Big Bang. But scientists now believe there was a moment when no light could travel through space because space was opaque. Photons just found themselves buffeted between one particle and the next. It was only 378,000 years after the Big Bang that the density of particles dropped sufficiently for the first photons to start their uninterrupted journey through space. This is when space suddenly had enough room for these photons to zip through the universe without running into something that might absorb them. These first visible photons of light make up what we call the cosmic microwave background radiation, and they represent the farthest that we can see into space. They are like a cosmic fossil telling us about the early universe.

Those first photons were only forty-two million light years away from Earth when they started their journey. Today, the distance has stretched to an estimated 45.7 billion light years. This is the edge of the visible universe, the cosmic horizon. But light isn't everything.

Although light couldn't make it through the plasma of space for 378,000 years after the Big Bang, neutrinos could. They are particles that don't seem to be stopped by anything (well, hardly anything; every now and again they bump into stuff that allows us to detect their existence). You have trillions of neutrinos passing undetected through your body every second. So it's possible that we could "see" a bit farther into space if we could detect the neutrinos that decoupled two seconds after the Big Bang.

There is a sphere surrounding the Earth that represents a horizon beyond which we cannot explore even with the most sophisticated of telescopes, because no information has had time to reach us. As time goes on, this cosmic horizon is growing, allowing us to see farther and farther into space. At least that's what we thought for a long time, but a discovery in 1998 revealed the alarming fact that, rather than extending farther into space, our cosmic horizons are actually contracting. Although the cosmic horizon is growing at a constant rate, the underlying fabric of space itself isn't just expanding, the expansion seems to be accelerating. As it does so, it is pushing things we can see out beyond our horizon, with devastating implications for what future generations can ever know.

THE STARS ARE GOING OUT

Some stars end their lives in a catastrophic explosion called a supernova. The luminosity is so intense that these supernovae can be seen over huge distances. When type 1a supernovae explode they all have the same luminosity, no matter where they are in the universe. When that is compared with the apparent brightness, we have a measure of how far away they are.

If the universe were expanding at a constant rate, then, given the distance of a supernova from Earth, you could predict the amount the wavelength of light should have been stretched by this constant expansion, or what we call its redshift. But when this theoretical redshift was checked against the recorded redshift of distant supernovae, astronomers got a shock. They didn't match. It should have been much higher. For galaxies farther away, whose light allows us to look back in time, the rate of change of redshift was slower than that of closer galaxies. The only possible conclusion was that the expansion of space in the early universe was much slower, but then it started to accelerate, ripping space apart.

It seems that something dramatic happened about seven billion years ago. Up to this point, the universe's expansion appeared to be slowing down, as one would expect as the gravitational force of the matter exerted a braking effect. But at this point, halfway through the current life span of the universe, the expansion rate changed character and started to increase, accelerating as if something had suddenly put its foot on the pedal. The fuel driving the acceleration is what scientists call dark energy.

It seems that in the first half of the universe's existence, the density of matter was enough to exert a slowing gravitational pull, but as the universe expanded this density decreased to such a point that the underlying dark energy was strong enough to take over. Dark energy isn't thought to be something whose density decreases with expansion. It is a property of space itself.

The sphere containing the visible universe is growing, which should mean that we can see even farther into space. Unfortunately, space is expanding so fast that stars that were previously within the sphere of the visible universe are being pushed beyond its edge. This suggests that in the future all galaxies other than our own will disappear from view and forever remain beyond the edge of our visible universe. Even though the sphere containing the visible universe is expanding, it will never expand fast enough to catch up with the galaxies as they are carried away by the acceleration of space.

Imagine if life had taken longer to evolve and humans had started doing astronomy only after all this interesting stuff had been pushed over the horizon. We would have an entirely different story of the evolution of the universe. It would look like the static universe that we believed to be our home before telescopes were able to pick out other galaxies. So what we can know depends on when we live.

Astronomy in the distant future won't consist of going to the top of mountains and peering through telescopes like the one at the Sphinx observatory. Instead, it will consist of consulting books and journals full of the data recorded by previous generations of astronomers, before what they observed was pushed beyond our cosmic horizon. Perhaps future astronomy will be more suited to a Thames Valley boy like me who prefers low-lying libraries to high-altitude observatories. But it will take another 150 billion years before the other galaxies disappear over the cosmic horizon.

It is worth noting that we won't ever lose the stars in our own galaxy. The local pull of gravity will keep the galaxy together. The expansion of space is not sufficient to pull the stars apart, but it raises the question of how much has already disappeared from view that might have told a different story.

If you are in a car and you want to accelerate, you need to put your foot on the pedal and produce energy by burning fuel. So where is the fuel or energy coming from that drives the acceleration of the universe, and won't it run out eventually?

The answer is that we don't know. *Dark* is used in cosmology to indicate that something doesn't seem to interact with light or other forms of electromagnetic radiation. In other words, we can't detect it. There are several conjectures as to what dark energy might be. One involves the cosmological constant that Einstein famously inserted into his equations to try to make the universe static. But now that constant is being used to push space apart. Usually we would think of energy spread out across space as being used up or thinning out as space expands. But this energy is now thought of as a property of space itself. More gets created as space grows. In any cubic meter of space it is constant. In other words, it has a fixed density. The acceleration is a runaway process that can't be stopped. This doesn't contradict the theory of the conservation of energy because dark energy is treated as a negative energy that is balanced by an increase in kinetic energy as space expands.

Information travels at the speed of light. With a static universe, this means that all information would reach us given enough time. With a universe expanding at a constant rate, my example of the ant on the expanding rubber band shows again that information traveling across an infinite universe would eventually reach us. But with an accelerating expansion, some things will never cover the intervening space quickly enough to counter the expansion. Using our current estimate for the cosmological constant we think might be responsible for this expansion, we believe that the sphere from beyond which we will not receive any information is currently eighteen billion light years in radius. As the space between the stars stretches, their light will stretch, too, and the more the light stretches, the longer the wavelength. Stars will seem to go out because the wavelength of light will be so elongated that we can no longer detect it.

It is amazing to think that cosmologists in the future may have no evidence that we live in an expanding universe. Future civilizations will perhaps return to the model of the universe held by the ancient world: our local galaxy surrounded by the void. We would look like an exceptional point in a universe of nothingness.

COSMIC FINGERPRINTS

If the universe is infinite, it seems very probable that its structure will preclude us from ever observing such a fact. And yet the universe beyond our visible horizon can still leave its fingerprints on the space we can see. If, on the other hand, the universe is finite, its shape will impose certain limitations on its behavior.

Think of the universe as a large resonating box like the body of my cello. The shape of the cello has been chosen because of the pleasing sound made by the frequencies that can vibrate within the box. What distinguishes a Stradivarius from a factory-made cello is partly the perfection of shape, which leads to a more beautiful sound. Can you deduce the shape of the box from the frequencies of the waves vibrating inside it? This is one of the intriguing problems that challenged mathematicians for some time. In a seminal paper, Mark Kac posed the question: "Can you hear the shape of a drum?" Only a square has the particular set of frequencies that are produced by a square. But in 1992, mathematicians Carolyn Gordon, David Well, and Scott Wolpert constructed two strange shapes whose resonant frequencies were identical, even though the underlying shapes differed.

In the mid-1990s, Jean-Pierre Luminet and his colleagues in France explored the cosmic microwave background radiation to see what waves were left over from the Big Bang. They believed waves with long wavelengths seemed to be missing from the spectrum. Was space not big enough to support them? Alas, more refined data from the Planck spacecraft released in 2013 has revealed that there is no such evidence of missing wavelengths that would have hinted at a finite universe. So the jury is still out. And therein lies the real dilemma—if the universe is finite, we could conceivably someday know it, but if it is infinite, we will forever be left in a state of epistemological unease.

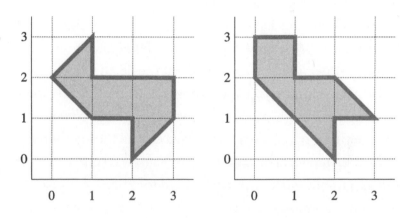

Two drums with the same resonant frequencies

Although the waves detected in the microwave background radiation can't tell us whether the universe is finite, they may allow us to estimate its minimum size. For starters, they give us a chance to sneak a peek beyond our cosmic horizon. As Soviet scientists Leonid Grishchuk and Yakov Zel'dovich explained in a paper in 1978, certain waves can resonate only if the universe is big enough. Patricía Castro, Marian Douspis, and Pedro Ferreira have used the resonances that we can detect to propose that the universe is at least 3,900 times bigger than the space we can see.

Another way we may be able to deduce things about space beyond the edge of our visible universe is to witness events that can only be explained by influences outside our cosmic horizon. For example, something large beyond our cosmic horizon might be pulling on the galaxies we can see, causing an unusual drift in certain regions of the night sky. Although we might not be able to see this large thing, we can still experience its effect on the things we can see. What we can know is not limited to what we can see. This is, in fact, how we know about dark matter. The gravitational behavior of the things we see makes sense only if there is more stuff out there. It is also how we discovered Neptune. Although we eventually saw Neptune with our eyes, its existence was initially predicted on the basis of the effect it was having on the planets around it.

As long as humans have been looking out at the night sky, from one generation to the next we've constantly found ourselves adjusting our sense of scale. At first it appeared that the Earth was the center of it all. Then we had to readjust our position in the cosmos when we realized that the sun was at the center, with the Earth just one of many planets orbiting the sun. Then we realized that all those stars out there might have their own planets and that our sun was tucked away at the edge of a galaxy of stars. And then we had to rethink our place in the universe once more as we came to terms with the fact that there were billions of other galaxies out there in space.

Discoveries made in my lifetime—since I'd come to terms with my place in the universe—have forced me to expand my horizons once again. Just as previous generations recognized that our planet was actually one of many, so it appears we have to face the possibility that our universe is simply one of many universes. The first inkling of these other universes was provided by a rather curious enigma encoded in the very first photons that zipped through our own universe.

THE MULTIVERSE

There is something curious about the cosmic microwave background radiation. Why is it so uniform? The photons that we are picking up in our detectors all have a temperature of 2.725 degrees above absolute zero. When they set off on their journey some 370,000 years after the Big Bang, they were much hotter—around three thousand degrees, the temperature at which electrons and nuclei amalgamate to become atoms. As space expanded, the photons cooled off, which means that their energy gradually decreased, until today it is so low that we find them in the microwave region of the electromagnetic spectrum.

But why do all photons have very nearly the same temperature? If two objects of different temperatures come into contact, then over time they will transfer energy from one to the other until their temperatures equal out. This may seem like the obvious explanation, except that there is an important word in that sentence: contact. Einstein's theory of relativity implies that information cannot travel faster than the speed of light. For one photon to communicate its temperature to another photon it needs to be close for long enough for this information to travel between them. This means that to reach a common temperature, opposite sides of the universe must have been much closer to each other much longer than our model of an expanding universe had suggested.

In the early 1980s, American cosmologist Alan Guth came up with a potential solution. He posited that in the early stages of the universe's existence, space didn't expand quickly. It was a slow start, allowing its photons to achieve a uniform temperature. Then space went through an exceedingly rapid expansion in a period that is now called inflation. An antigravity field confusingly called the inflaton is conjectured to have pushed space apart at an exponential rate. The rapid inflation didn't last for long: some 10^{-36} seconds, according to the current model. That's a billionth of a billionth of a billionth of a billionth of a second. Yet in that time, space is thought to have expanded by a factor of 10^{78}. It's as if a buildup in pressure had suddenly been released, after which the universe settled down into the more sedate expansion we have picked up since.

This model helps explain why the universe appears flat and essentially homogeneous. The large variations we see, like galaxies here but empty space there, are actually the product of very small quantum fluctuations in the small bit of space that were blown up by this massive inflation. The inflation also explains why the universe appears so flat. It would have virtually flattened out any identifiable curvature that had been present in the early universe.

The mathematics developed to explain this inflation, by Andrei Linde at Stanford and Alexander Vilenkin at Tufts, makes an extraordinary prediction. It implies that space could be inflating in other regions of space. Quantum fluctuations across space cause the inflaton field to pop in certain places, creating vast universes. In other words, there may be other universes out there like ours, making space look like Swiss cheese, with the holes in the cheese corresponding to different universes.

Can we ever know whether this description of the universe is true? Are we just coming up with internally consistent stories that could be true but are untestable? Even in our own universe there seems to be a limit to how far we can see. So how can we hope to know whether these other universes are real or just the fantasy of theoretical physicists?

Scientists hoping to prove the existence of these other universes have focused on the possibility that they might interact with our universe in such a way as to leave some sort of imprint. Could the cosmic microwave background radiation provide evidence that our universe collided with other universes during its formation? There has been some speculation that temperature differences in the map of the early universe could be a result of such collisions, but, as a member of the team at University College London investigating this possibility admits, "One of many dilemmas facing physicists is that humans are very good at cherry-picking patterns in the data that may just be coincidence." It is devilishly complicated, but not a hopeless quest. Discovering evidence for neighboring universes is not *a priori* something beyond our reach. The challenge is to find ways in which these other universes could have influenced what we see within our cosmic horizon.

DIALING UP DIFFERENT UNIVERSES

One of the intriguing consequences of these potentially unknowable universes is that they provide us with a way to answer one of the principal questions with which many believers challenge those who don't believe in a God. The possibility of the multiverse offers the best rationalization we have so far for the troubling feeling that our own local universe must have been designed.

I'm not talking about the illusion that only a supernatural designer could explain the complexity of life. Thanks to Darwin's theory of evolution, we don't need a designer. But what we lack is a good explanation for the constants of nature—the twenty or so numbers like the mass of an electron, the gravitational constant, the speed of light, the charge of a proton—that had to be so perfectly fine-tuned to get life up and running. Although we have a mechanism that explains biology, we don't have a similar way to explain physics.

There seems to be no clear rationale for why these constants assume the values they do. Why couldn't they be dialed to other values? What is particularly striking is how sensitive the possibility of life in our universe is to a small change in these constants. For example, if the constant that controls the way the electromagnetic field behaves is changed by four percent, then fusion in stars could not produce carbon—any life that could exist in such a universe would have to be based on other atoms. Some of the other constants turn out to be similarly sensitive to small tweaks. Change the cosmological constant in the 123rd decimal place and suddenly it's impossible to have a habitable galaxy.

The multiple universes beyond our cosmic horizon could hold the solution to this dilemma. In the multiverse model there are lots of different universes, and in each of these universes the fundamental constants could be randomly assigned. In most cases, a universe wouldn't really have much going on because the constants aren't set favorably for action. But in a few, the constants are in the sweet spot for atoms to form out of the soup of electrons and quarks in a way that eventually leads to life. Of course, for us to be around to observe this universe, we have to be in one of these special universes. This is known as the anthropic view.

I think most scientists hope for a more satisfying answer to why our universe is the way it is. They'd like to be able to show that these constants aren't really random, that things had to be the way they are. Relying on multiverses feels like a bit of a cop-out. But perhaps I have to bite the bullet and accept the multiverse model. The Earth may just be a random planet that turns out to be in the sweet spot for life. I'm not going to be able to come up with a reason why it had to be there. Lots of planets failed to be in the right spot. It's just chance that our planet fits the bill. Perhaps our universe is the same, one of many different universes that just happens to be in the sweet spot for life.

My hope is that we can explain why the universe has to be the way it is without recourse to the idea of the multiverse. That is what most scientists would prefer. Some suggest that there is an explanation—a transcendental intelligence that fine-tuned the whole thing. But that feels like even more of a cop-out. The multiverse theory at least has that sense of economy we are after in a good theory. The explanation terminates and does not require further explanation. The addition of these other universes is just more of the same with variations, and once you accept all these universes you get a complete solution to the fine-tuning problem. A designer who fine-tunes the constants raises as many questions as it answers.

A good scientific theory should make sense of how everything is put together, and you shouldn't need to introduce too many extra characters into the story to get the narrative we experience. There is a simplicity and naturalness about the multiverse theory that makes it a strong candidate theory. The physics of inflation provides a possible mechanism for producing these other universes, so it isn't just a wild hypothesis. But we need to take care when using the qualities of simplicity and economy to judge the possible truth of a theory.

If I asked you to name the next number in the sequence 1, 2, 4, 8, 16, . . . , the obvious answer would be 32. To guess that doubling is the key to this sequence would be the majority view. But what would you say to the person who announces that 31 is the next number? That person would probably be laughed at. But once you know that these numbers could describe the number of ways you can divide a circle, both 31 and 32 become legitimate responses. So then you may ask for more experimental data to help you decide which is the best explanation.

Why 31 Is the Next Number . . .

Place n dots on a circle and then draw lines connecting each dot to the other dots. How many regions have you divided the circle into? The maximum number of regions you get starting with one dot follows the following sequence: 1, 2, 4, 8, 16, and then at six dots there is a surprise—you get a maximum of thirty-one regions.

The interesting point is that, however much data you uncover, I can give you an equation that will provide a coherent reason for why any number might legitimately come up next. So, given a finite amount of data, it seems we can never really know what the explanation is for that data unless we can get another bit of data against which to test our equation. This is the model of science that the philosopher Karl Popper proposed: a theory can only be falsified, never proved.

In judging a good theory, scientists often apply some measure of simplicity, an Occam's razor measure that favors one equation over another. The simplicity of the equation—that is, the less you have to put in to get an answer out—is generally significant when deciding which explanation to favor. That's why most would choose doubling as the obvious explanation for the sequence of numbers, rather than the quartic polynomial that gives us 31.

So, when we are faced with two competing theories and there is no way of telling which one is right, it does seem to be the case that we favor the simplest theory. This is captured by the idea of something philosophers call "inference to the best explanation," or the theory of abduction. But there is no reason why simplicity should guarantee truth.

Our association of beauty with truth comes from our being evolutionarily programmed to respond with a shot of dopamine when we think we've struck on an idea that helps us to navigate our environment. We call things beautiful because of our body's response to something that will be advantageous to our evolutionary survival.

What happens when alternative stories are proposed? How do you deal with people who believe the universe is 5,775 years old? You show them the fossil record and they counter that the universe was created old. They construct for themselves a narrative that is logical and self-consistent but to my mind highly unlikely. It is hard to argue with someone who posits a theory that is immune to testing.

Increasingly we are being pushed toward scientific narratives for the evolution of the universe that may turn out to be untestable. If you come up with a theory that predicts new particles but the theory doesn't pin down at what energy these particles can be detected, no amount of evidence will convince those who believe in this new theory that they are wrong. They will always retort that the hypothetical particles exist in the region that you haven't been able to test yet.

Some have argued that the current untestability of the multiverse theory means that it is as fanciful as proposing that a supernatural designer fine-tuned the whole thing. Although at the moment there is no way of testing the multiverse theory, there is no *a priori* reason why it will always remain untestable. The same applies to string theory, which is often shot down as a scientific theory because it fails to make predictions that can be tested. But that is no justification for throwing it out yet, as there is no reason to believe that it will always remain untestable.

The multiverse theory, although potentially untestable, does come with a mechanism, inflation, for how these multiverses arise. And we do at least have evidence for one of these multiverses: our own universe. One of the criteria for a scientific theory is that it should provide explanations based on things that are natural rather than supernatural. If you are going to posit a new thing like "dark energy" or "gravity," you need to embed it in the natural world. How can you do that? You show how its behavior affects the rest of what we see and experience.

Another criterion for good science is that you can do experiments to test your theory. One of the problems with cosmology is that it is a one-time experiment. It is very difficult to run another Big Bang and to see what happens. That said, you can test the conditions of the Big Bang on a small scale on Earth to get a sense of the physics of the larger cosmological event. But it's difficult to see how you could test whether other physical theories with different fundamental constants, or even different physics, could emerge from different moments of creation.

Since experiments are out, one of the central tenets of cosmology is the concept of homogeneity. You have to work under the assumption that what happens in our local bit of the universe is representative of its overall structure. Without this assumption, anything goes. We assume that the way space curves near us is true universally, but this need not be the case, just as someone living on the flat part of a hemisphere-shaped planet will think the whole planet is flat until they reach the point where curvature changes.

Without the assumption of homogeneity, how can we preclude the possibility of something outrageously different going on beyond our horizon? Perhaps someone has downloaded our universe from a website and glued it together like my celestial sphere. Perhaps we are just the dollhouse of some supernatural being—supernatural in the sense that it exists outside our universe. If this supernatural being never plays with the dollhouse, it is unclear that we could ever know about it. But if it truly has no influence, then why imagine that it exists? Why do we need to let our imaginations run riot? If this supernatural being plays with its dollhouse, then we are up for testing its impact and it is potentially knowable.

IS THERE ANYONE OUT THERE?

It is striking how often cosmology and religion have intersected throughout history. The question of what might lie beyond the edge of the universe has always intrigued scientists and theologians alike. For the medieval philosopher Oresme, this was where God was hiding. Most religions include a creation myth for the cosmos. For the Australian Aborigines everything spewed forth from the belly of the Rainbow Serpent. For scientists it's the Big Bang. Galileo ran into trouble with the Catholic Church because he challenged its conception of our place in the cosmos. Yet it was a Jesuit priest, Lemaître, who came up with science's current story of how our cosmos emerged from the Big Bang.

Even in our current era, cosmology and religion are sometimes controversially intersecting. In 1972, Sir John Templeton, an American-born British entrepreneur, introduced a prize (given in his name) to reward "progress in religion." It has evolved to recognize "progress toward research or discoveries about spiritual realities." It is a big-money prize, currently standing at £1,100,000. Templeton stipulated that the prize should exceed the Nobel Prizes, which he felt had unjustly ignored the spiritual dimension.

The prize has been awarded to some obvious people. The first Templeton Prize went to Mother Teresa. Subsequent prizes have rewarded priests, evangelists, rabbis, and the Dalai Lama. But in recent years a growing number of scientists have joined their ranks, and invariably it's scientists researching cosmology and big questions about the universe who have been honored.

A number of leading scientists have been critical of those who have accepted the prize, believing that it endorses a spiritual approach to scientific questions. My predecessor Richard Dawkins, one of those critics, declared that the prize was usually given "to a scientist who is prepared to say something nice about religion." Physicist Sean Carrol outlined his reasons for refusing funding from the Templeton Foundation. "It's not a matter of ethical compromise," he wrote. "It's simply a matter of sending the wrong message. Anytime respectable scientists take money from Templeton, they lend their respectability—even if only implicitly—to the idea that science and religion are just different paths to the same ultimate truth."

I was keen to talk to a cosmologist who had accepted the prize and approached Professor John Barrow, based in the Department of Applied Mathematics and Theoretical Physics at the University of Cambridge, who received the Templeton Prize in 2006. "Most of the fundamental questions in cosmology are unanswerable," he immediately wrote back in response to my e-mail. "In fact, I will talk about some of them in a lecture I'm giving this Saturday."

Perfect. Here was a chance to hear what cosmologists believe are the great unknowns. Sitting in the front row of the lecture hall, I waited impatiently for him to begin. As Barrow went from one question to the next, accompanied by the obligatory pretty pictures with which astronomers pepper their presentations, he proceeded to confirm all the fears I've been accumulating about how little we can know about our universe.

The Big Bang: "At a finite time in the past the universe seemed to have a beginning—a time when the density was infinite and the temperature was infinite. We don't know if that beginning was real."

The size of the universe: "There is no doubt a lot of universe beyond our horizon, but we haven't seen it, and almost all of it we will never see. So when people ask whether the universe had a beginning, or is it finite or infinite, we can never answer questions like those about the entire universe."

When I met up with Barrow after the lecture and challenged him about the Templeton Prize, he explained to me why he thought the Templeton Foundation was funding and rewarding cosmologists. "Sir John Templeton liked science because there was progress. He believed cosmology was tackling deep and important questions that everyone ought to know about whatever they were doing, whether it be philosophy or religion or theology. You couldn't study those subjects while ignoring what was going on in the rest of science."

Barrow believes the science–religion debate needs to be much more nuanced. "One of the lessons to be learned for people who are interested in the science–religion interaction is that you have to say which science you mean because the interaction is very different.

"Cosmology and fundamental physics bring you into contact with these big questions that you know you're not going to answer. You're used to uncertainty, to not knowing, and, to some extent, seeing why you're not knowing. People in lab physics or biology, like Dawkins, they're not used to that kind of situation. They think every problem is solvable by grinding it down."

Not that Barrow was belittling what they do. "People always think that because you're studying the universe this must be the hardest problem. Not at all. Understanding the brain, for example, is a much harder subject. It's much more complicated. In cosmology things happen very slowly, there are nice approximations for understanding. You can pick a simple symmetrical solution and then iterate away. You can't do that with human society. There's no simple model."

Barrow believes that the science he is engaged in has a markedly different quality. "In fundamental physics and cosmology you know there are problems that you're not going to solve. Why are there laws of nature at all? The question of the number of dimensions of space and time. Whether there is a multiverse. Was there an initial singularity? Is the universe infinite? You can think of loads of questions that you're not going to answer. This creates a different complexion."

Barrow received the Templeton Prize for his work highlighting the intrinsic limitations of scientific inquiry. There is a growing tendency to think that science will ultimately be able to fill all the gaps, and Barrow is keen to rein in this faith in the all-knowing power of science. He is a big believer in the Copernican principle that we are not at the center of things and that the universe does not revolve around the human race.

"The universe is not constructed for our convenience. It's not an exercise in the philosophy of science. It's too bad if we can't find these things out. In fact, I'd be very suspicious if all these fundamental questions happened to be answerable by what we're doing. I would regard that as anti-Copernican. So I regard the fact that we can't solve certain problems, or we can't get the data we need, as a Copernican aspect of things."

BLACK SWANS, BIASES, AND THE BACK OF THE BOOK

Barrow believes we must recognize that we have an extraordinarily biased view of our universe. "Most astronomy is based on observing things that shine in the dark: stars in distant galaxies, so-called luminous matter. Only five percent of the universe is composed of ordinary material that composes us and the luminous stars. Luminous material is rather biased. Luminous material tells you about places in the universe where the density became so high that nuclear reactions could ignite and produce luminosity."

In some ways, this applies to the whole of science. Our view of the universe is biased toward the bits that have impact on our senses. Anything that doesn't isn't going to get noticed.

"If you lived on a planet that was covered in cloud all the time—planet Manchester, say—then there would be no astronomy. On the other hand, you might learn an awful lot about meteorology."

In Barrow's view, cosmology has a very different quality to other sciences, and this may be key to its close relationship with theology. "If you're a scientist you're used to carrying out experiments, testing theories. You can't experiment on the universe. You just have to take it as it is."

Many modern scientists have signed up to Karl Popper's conception of the limits of knowledge—that because you can never actually prove that a scientific theory is correct, the best you can do is to try to falsify it. All swans are white. You can never prove that. The best you can do is to falsify it by discovering a black swan. In Popper's book, anything that can't potentially be falsified is therefore unscientific. So if you can't do experiments in cosmology, are large swathes of it unscientific? Barrow wasn't going to give in too quickly.

"Popper's philosophy of science is incredibly naive. It doesn't really work in astronomy, because when you make your observations you don't know whether they have been made correctly. So your prediction might have been falsified because something has been done wrong in the experiment. Or, more likely, there is some bias in the way the evidence has been gained."

Black swans seem pretty damning for a theory that says all swans are white, but in other more complex settings, it might be much less clear whether the theory is at fault, or the evidence. Perhaps the swan rolled in coal dust before you observed it.

Barrow believes that cosmology could have its own black-swan moment. "The key data point in all these ultimate questions that could really change things is making contact with some well-evolved extraterrestrial civilization that has evolved independently and knowing how they have come to think about some of these questions. Have they found the same bits of math useful? Have they done physics in the same way? Do they define fundamental constants in ways that we could make sense of? What do they think about these ultimate questions? Now this would be an unsurpassably important data point and is the most important reason to make contact with extraterrestrials."

Given the tantalizing possibility that this advanced civilization might have come up with answers to some of our big questions, I thought Barrow would be excited by the possibility of contact. But I got a surprise.

"The scientific content I think will be a disaster. If we made contact with a very advanced civilization that had answered all the questions we had ever posed, the game would be over for us. There would be no motivation for us to do science. It would be like looking at the back of the book and finding all the answers."

You wouldn't want to do that?

"I think it would be a disaster."

Just as particle physicist Melissa Franklin didn't want to press the button that would give her complete knowledge, so Barrow was averse to looking up the answers in the back of the book. Unlike Franklin, Barrow is convinced that there are questions that even the most advanced civilization will never be able to answer. Those pages, to his mind, will forever remain blank.

"The bad news is that cosmology has unknowable unknowns: things that we are never going to know even though we suspect we might know of their existence."

MAKING CHOICES

Many would argue that you should remain agnostic in the light of an unknowable question, so I was intrigued to know Barrow's stance on the question of God. Was he agnostic? Atheist?

"Christian, in fact."

That was a surprise. I have heard Barrow talk many times, and have read many of his books, but unlike Polkinghorne, he does not wear his beliefs on his sleeve (or around his neck). But I guess I also have made a choice. I don't believe that the only logical response to an unanswerable question is to remain on the fence.

Our response to an unknowable question can have implications for how we conduct ourselves. Take, for example, the question of whether the universe is infinite. That may well be a question that we will never be able to answer. So should we just remain agnostic on the subject?

There is an argument that we would be wise to take out Pascal's wager on this one. If the universe is infinite, you're (probably) never going to know. But if the universe is finite, it's possible that you could know this. So isn't the better strategy to choose to believe that the universe is finite? In the end, if the universe is infinite you can never be proved wrong, but if it's finite you could be proved right.

But what if an infinite universe offers a more exciting framework within which to live your life? An infinite universe has some interesting implications. One of these is that there could be infinitely many copies of you across the universe reading this book. The psychological impact of this conclusion might have a dramatic effect on the way you live your life.

Why There Could Be Infinitely Many Copies of You Reading This Book

This hinges on a couple of assumptions. The first comes from quantum physics; namely, that everything in the universe is quantized. This means that in a finite region of space there are only finitely many points and they can take only a finite number of different values.

A simplified version of this universe looks like an infinite checkerboard on which each square can only be black or white. Take a region of this checkerboard, which is meant to represent the complexity of life, including you reading this book. Suppose, for example, that it is a 10 x 10 region with a particular pattern of black and white squares.

One model of the entire infinite universe is that outside this 10 x 10 region everything else is colored black. Rather like the void. So we need another assumption: that it is equally probable that any of the possible patterns can occur. No pattern is favored over any other. Since there are only finitely many possible patterns and infinitely many 10 x 10 cells across this universe, if our pattern occurred only finitely many times then there must be another pattern that occurs infinitely often and therefore is infinitely more likely, which contradicts our second assumption. That means our pattern must be repeated infinitely across the checkerboard universe.

I think this is why I reject Pascal's original wager on the question of the existence of God (in the conventional sense of a supernatural intelligence that created the universe) and choose to declare myself an atheist. Ultimately it affects the way I conduct my life. I'm not denying that it is a question I may never know the answer to, but to admit such leaps of the imagination would open up too many other wild possibilities. It goes against my natural predisposition to argue for the best explanation. Believing in multiverses may not be the most satisfying thing, but it is to my mind a better—or simpler—explanation.

CAN WE REALLY EVER KNOW WE WON'T KNOW?

After leaving Barrow in his office, I was rather downcast. I know I was on the lookout for things we cannot know, but I was beginning to doubt whether we could actually ever know anything. On my way back home, lines in one of Barrow's books struck a chord: "The idea of the impossible rings alarm bells in the minds of many. To some, any suggestion that there might be limits to the scope of human understanding of the Universe or to scientific progress is a dangerous meme that undermines confidence in the scientific enterprise."

Looking back at the Edges we've visited so far, nothing seems quite so unanswerable as the question of whether the universe is infinite. Chaos theory told me that the future was unknowable, but I can just wait until it becomes the present and then I'll know. As I slice my die, I might actually hit a point at which space is quantized, so it's possible that there are only finitely many steps inside my die before I hit the indivisible. It's true that it might be almost impossible to navigate this parsing into even smaller pieces, but it isn't a challenge that is *a priori* unsolvable. And Heisenberg's uncertainty principle does not so much challenge us to give an answer as to consider whether the questions we are asking are well posed. It's not that we can't know position and momentum simultaneously; the revelation is that it is not a question it makes sense to ask.

But the question of an infinite universe does not seem ill posed. The universe is either infinite or not. If it is infinite, it is a real challenge to think of how we could ever know this, given what we've discovered about the cosmic horizon.

Then I had a revelation. Perhaps the question of an infinite universe is not as unknowable as one might think. Might there not be a more indirect way to reach a conclusion? The answer might lie in my own field of expertise. Mathematics has been a very powerful telescope through which to view the universe. What if the current laws of physics lead to a mathematical contradiction under the assumption that the universe is finite? That would force us to conclude that the universe must be infinite, or that our laws of physics are at fault. This is, after all, how we discovered irrational numbers.

This is in fact the power of mathematics: to use our finite brains to know the infinite. The Pythagoreans showed that the length across the diagonal of a unit square is given by a number that can't be written as a simple ratio of whole numbers. The only way this length can be expressed is by a decimal expansion that is infinite and nonrepeating. Perhaps an infinite universe will be susceptible to a proof of existence by the same tools that helped us discover irrational numbers—a proof by contradiction.

Perhaps the real lesson is that "what we cannot know" is something we can never know because it is so hard to preclude the possibility that new ideas might pull the unknowns into the known. Although the universe that I will ever be able to see or explore is as finite as the paper model that sits on my desk, perhaps the message is that we should not give in too early to the lure of the great unknown. The mathematical telescopes of the mind may one day allow us to break through the unknown and discover that our sphere really does sit in an infinite expanse of space.

FIFTH EDGE

TIME

9

No people who have the same word for yesterday and tomorrow can be said to have a firm grip on time.

—Salman Rushdie, *Midnight's Children*

I T'S 8:50 . . . ROUGHLY . . . or so my wristwatch tells me. The weak February sun is trying to climb above the rooftops of the houses opposite. The radio is on, the coffee is brewed. The beginning of another day. But the pulse of Prokofiev's *Cinderella* vibrating the speakers alarmingly reminds me that time is pressing on. The chimes are telling Cinderella that her time at the ball is up. And here I am procrastinating on the Internet. I've just put my birthday into Wolfram Alpha and it's telling me that I've been alive for 18,075 days. But when I ask it how many more I have left, it says it doesn't understand my query. Probably just as well. I'm not sure I want to know how many more times the hands on my watch will go around before the pulse beneath stops.

When I was younger I thought it might be possible to know everything. I just needed enough time. As the years tick by, I'm beginning to realize that time is running out. That youthful sense of infinity is giving way to the middle-aged acknowledgment of finitude. I may not be able to know everything. But is there still hope that humanity can know it all? Or will time run out for us collectively too? Whether space is infinite may well be something we can never know, but what about time? I think we all have the feeling that time will probably go on forever. Keep putting batteries in my watch and it will carry on ticking. But when it comes to the question of when it all began, people are less certain: Did time have a beginning, or has it been there forever?

I may not be able to look into the future and make predictions, but the past has happened, so shouldn't I be able to look back and see if time had a beginning? Tracing the expansion of the universe backward has led us to a moment we call the Big Bang, when space was infinitely dense, a singularity that occurred 13.8 billion years ago. But what about before the Big Bang? Is that simply a no-go area for scientific investigation? Or are there telltale signs in the current state of the universe that might show me what was going on before it all began?

The nature of time has troubled philosophers and scientists for generations. Tied up in the attempt to understand this slippery concept is the challenge of understanding why there is something rather than nothing. To talk about a moment of creation is to talk about a moment in time.

Einstein's theory of relativity—with its revelation that time and space are inseparably linked—seems to suggest that time exists only once you've created space. But if time and space came into being only at the Big Bang, the concept of time "before" the Big Bang has no meaning.

There are rumblings in the cosmological corridors. Perhaps time can't be mathematically packaged up so easily. Perhaps the question of what happened before the Big Bang is not so easily dismissed. But to try to unravel time means diving into some pretty mind-bending ideas.

As I stare at my watch I can't actually see the hands moving, but if I look away and then look back again after some time has passed, the hands have moved on. It's saying 9:15 now ... or thereabouts. Tiny cogs inside the watch are driven by a tiny electric motor that in turn is driven by one-second pulses that have their origin in the oscillations of a tiny quartz crystal. The battery that sits inside the watch sets up a voltage across the crystal that makes it vibrate like a bell with a frequency tuned to 32,768 vibrations per second. This number is chosen for its mathematical properties. It is 2 to the power of 15. Digital technology likes powers of 2 because computer circuitry can quickly convert this into a mechanical pulse to drive the cogs every second. Importantly, this frequency is not greatly affected by the surrounding temperature, air pressure, or altitude (factors that affect a pendulum, for example). And this vibrating, the repetition of motion, is key to marking the passage of time. But is this sufficient to describe the idea of time?

With my watch ticking away on my wrist, there are no further excuses for procrastinating. So . . .

WHAT IS TIME?

Most attempts to define time very quickly run into difficulties that become quite circular. It's something my watch keeps track of . . . It's what stops everything from happening at once . . . The fourth-century theologian St. Augustine summed up the difficulty in his *Confessions*: "What then is time? If no one asks me, I know: if I wish to explain it to one that asketh, I know not."

Measuring time is, at heart, a very mathematical process. It depends on spotting things that repeat, things with patterns in them, such as day and night, the movement of the planets or the passing of the seasons or the swinging of a pendulum or the pulsing of an atom. As the nineteenth-century Austrian physicist Ernst Mach had it, "Time is an abstraction at which we arrive through the changes of things."

The Lascaux caves in France are believed to contain evidence of one of the first attempts by humans to keep track of time. Dating back fifteen thousand years, the caves were discovered in 1940 by four

French boys after their dog Robot came across a hole leading down into the rocks. The caves are famous for the extraordinary Paleolithic paintings of animals running across the walls: bison, horses, deer, and aurochs.

I've had a chance to visit the caves, but because of the delicate nature of the drawings, I got to see only the replica cave that has been built alongside the original. They are still impressively atmospheric, and there is a dramatic sense of energy captured by the ancient images. But it isn't only animals that the artist depicts. There are also strange arrangements of dots that punctuate the paintings, which some archeologists believe are evidence of ancient humans keeping track of the passage of time.

One collection of dots is generally believed to depict the constellation of the Pleiades. The reappearance of this constellation in the night sky was regarded by many ancient cultures as marking the beginning of the year. As I moved around the cave I came to a sequence of thirteen dots with a rectangle drawn at one end. Above the rectangle is a huge image of a rutting stag. Farther along the wall there is a sequence of what could be counted as twenty-six dots with a huge pregnant cow at the end.

Some archeologists have interpreted the dots as marking quarters of the moon's cycle, what would become the seven days of the week. These quarter phases of the moon were easily identifiable in the night sky. So thirteen quarters of the moon represents a quarter of a year, or a season. Counting on a quarter of a year from the reappearance of the Pleiades gets you to the season of rutting stags, when they are more easily hunted. Then the twenty-six dots can be interpreted as two lots of thirteen dots, representing two seasons, or half a year. This gets us to the point in the year when the bison are pregnant and again vulnerable and easily hunted.

The paintings on the wall might represent a training manual for new hunters, a calendar telling them what to hunt at what point in the annual cycle. This early evidence of timekeeping relies on spotting patterns that repeat themselves. Identifying repeating patterns would always be key to understanding the nature of time.

The cycle of the sun, moon, and stars would inform the way we measured time until 1967. There is nothing in the natural cycle that dictated how we partition our day. Rather, it was the mathematical sensibilities of the Babylonians and Egyptians that gave us a day divided into twenty-four units of time and an hour subsequently divided up into units of sixty. The choice of these numbers was based on the high divisibility of the numbers 60 and 24. Napoleon did attempt to make time decimal by introducing a day with ten hours, but it was about the only unit of measurement that he failed to get the world to count with their ten fingers.

Until 1967, the second, the basic unit for measuring time, was defined variously in terms of the time it takes the Earth to rotate on its axis or the Earth to rotate around the sun, neither particularly constant when measured against our modern concept of time. For example, six hundred million years ago the Earth rotated once on its axis every twenty-two hours and took four hundred days to orbit the sun. But the tides of the seas have the strange effect of transferring energy from the rotation of the Earth to the moon, which results in the Earth's rotation slowing down and the moon gradually moving away from us. Similar effects are causing the Earth and sun to drift apart, changing the time it takes to complete an orbit.

Given the vagaries of the motions of the planets, from 1967, instead of measuring the passage of time by looking outward to the universe, meteorologists looked to the atom to define the second, which is now understood as "the duration of 9,192,631,770 periods of the radiation corresponding to the transition between the two hyperfine levels of the ground state of the cesium-133 atom at rest at a temperature of 0 K."

Quite a mouthful. And you should see the clock that does the measuring. I visited the National Physical Laboratory in southwest London to see the atomic clock that tells Big Ben when to sound the hour. It's enormous—definitely not something that you could wear on your wrist. It includes six lasers that trap the atoms of cesium before launching them upward into a microwave chamber. As they fall back down under the effect of gravity in what's called a cesium fountain, the atoms are zapped with microwaves, causing them to emit the radiation whose frequencies are used to define the second.

The atomic clocks that sit at the national laboratories across the world are some of the most extraordinary measuring devices created by humans. The regularity and universality of the atom means that two atomic clocks put next to each other would, after 138 million years, differ by at most a second. These clocks produce some of the most precise measurements ever made by humans. So perhaps we *can* say we know time. The trouble is that time isn't as constant as we had hoped. If two atomic clocks are moving relative to each other, then, as Einstein famously revealed at the beginning of the twentieth century, they will soon be telling very different stories of time.

TORCHES ON TRAINS

Newton believed that time and space were absolutes against which we could measure how we move. In the *Principia* he laid out his convictions: "Absolute, true and mathematical time, of itself, and from its own nature flows equably without regard to anything external."

For Newton, space and time were like a backdrop upon which nature played out its story. Space was the stage on which the story of the universe was performed and time marked the passage through this story. He believed that you could place clocks across the universe, and once they had all been synced they would continue to show the same time. Others were not so certain. Newton's archrival Gottfried Leibniz believed that time existed only as a relative concept.

A discovery made in 1887 by Albert Michelson and Edward Morley ultimately led to Leibniz's view winning out over Newton's. The American scientists found that if we measure the speed of light in a vacuum, regardless of whether we are moving toward or away from the source of the light, the measurement remains the same. This revelation was the seed for Einstein's discovery that time was not quite as absolute as Newton had envisioned.

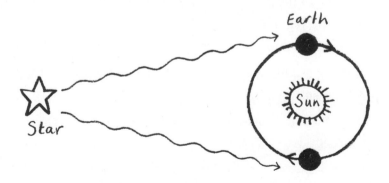

At first sight, the fact that the speed of light would be the same however I am moving relative to the light source seems counterintuitive. Consider the Earth orbiting the sun. If I measure the speed of the light coming from a distant star, I would expect it to be faster when I am heading toward the star than when I am heading away from it.

Newtonian physics implies that if I am running ten miles an hour down a train traveling at ninety miles an hour, then relative to someone on the platform I am traveling at a combined speed of a hundred miles an hour. So why isn't the same true of light shining from a torch on the train? Why isn't the speed of light ninety miles an hour faster for the person measuring it on the platform? It transpires that Newton was wrong about how to calculate the speed of the runner on the train relative to someone on the platform. You can't simply add the speed of the runner to the speed of the train. The calculation turns out to be more subtle.

Relativistic Formula for Combining Speeds

To calculate the apparent speed s of a passenger running at u miles an hour down the length of a train traveling at v miles an hour, Einstein discovered that you need to use the following equation:

$$s = \frac{u + v}{1 + (uv/c^2)}$$

where c is the speed of light. When the speeds u and v are small compared to c, the term uv/c^2 is very small. This means that the speed s can be approximated by adding the speeds $u + v$. But when the speeds u and v get close to the speed of light, the approximation breaks down and the formula gives a different answer. The formula is such that it never outputs a combined speed that is faster than the speed of light.

It was by trying to understand why the speed of light is invariant that Einstein, in 1905, made the breakthrough that changed our perception of the universe. He discovered that time and space are not absolute but vary according to the relative motion of the observer. Einstein was employed at the time as a clerk in the Swiss patent office and was responsible for evaluating applications for a range of new inventions, from gravel sorters to electrical typewriters. But he also had to assess attempts to create devices to electrically synchronize time, an important task for a world that was becoming increasingly interconnected. It was such apparently mundane work that fired the thought experiments that led to Einstein's special theory of relativity.

SLOWING DOWN TIME

To explain Einstein's new ideas about time, I need a clock, or something that repeats itself at regular intervals. I could use one of the atomic clocks at the National Physical Laboratory or my wristwatch, but, in fact, light is the best clock to use to reveal the strange effect of motion on time. I am going to exploit Michelson and Morley's discovery that the speed of light does not seem to depend on how fast you are moving when you measure it.

So let me consider time measured by a device consisting of two mirrors with a light beam bouncing between them. Each tick of this clock corresponds to the light bouncing between the two mirrors.

I am going to take one such clock onto a spaceship and the other I will leave with you on the surface of the Earth. To show that—from your perspective on Earth—my spaceship's clock ticks at a slower rate requires nothing more sophisticated than Pythagoras's theorem. That said, the argument is still quite subtle and will require some deep concentration to comprehend. This is, after all, one of the most extraordinary revelations of the twentieth century, so it isn't going to be

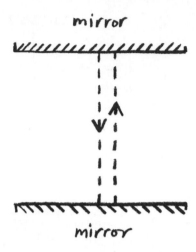

easy. The reward for the effort will be to change your perspective on time in quite an astounding manner.

Einstein's revelation depends on the important discovery made by Michelson and Morley that the speed of light is the same everywhere. This means that the person on Earth must record the light bouncing at the same speed between the mirrors in the clock on the spaceship as it does on Earth. Just because the spaceship is moving, that movement can't add speed to the light. So let us look at how far the light travels when it is emitted from one mirror and hits the opposite mirror in my clock on board the spaceship.

Suppose the mirrors in both clocks are four meters apart. Let us suppose also that in the time it takes for the light to travel between the mirrors on the spaceship, the ship has moved three meters as measured by you on Earth. Pythagoras's theorem tells us that the light has traveled across the hypotenuse of the triangle and has therefore covered five meters. That's all the math you need to understand Einstein's theory of special relativity!

From this we can calculate that the spaceship is traveling at 3/5 the speed of light relative to the Earth—in other words, the spaceship covers three meters in the time that it takes light to travel five meters.

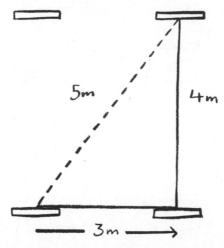

The spaceship's clock moves three meters in the time it takes the light to travel between the mirrors set at four meters apart. By Pythagoras's theorem, this means the light has actually traveled five meters.

The key point is that light on the Earth will have traveled the same distance—five meters—since the speed of light must be the same everywhere. But the distance between the mirrors in the clock on the Earth is also four meters. This means it must have hit the top mirror, and be on its way back down again, one quarter of the way through a second tick. So time is running faster according to the person on Earth because one tick of the clock on the spaceship takes the same time as 1¼ ticks of the clock on Earth. According to you, my clock in space is going ⅘ slower!

To get a sense of why this is happening, take a look at the beam of light as it travels on the spaceship and on Earth. The beam of light is going at the same speed in both clocks. The freeze-frames on the next page show where the light will be at various points in time. Because the light on the spaceship has to travel through space in the direction of the movement of the spaceship, it can't travel as far in the direction of the opposite mirror in the clock on the spaceship. So from your perspective on Earth, the light in your clock reaches the mirror before the light on my clock on the spaceship does. This means that your clock is "ticking" faster.

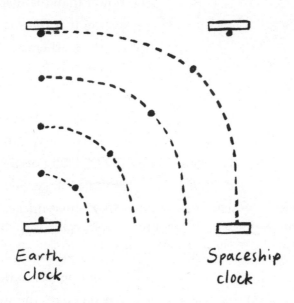

Earth
clock

Spaceship
clock

Fair enough . . . and yet when I compare the clocks from my perspective on the spaceship, things get very counterintuitive. To understand why things get so weird I need to apply what is known as the "principle of relativity." This says that if you are in uniform motion (i.e., not accelerating and not changing direction), then it is impossible to tell that you are moving. This principle of relativity can be found in Newton's *Principia*, although it is probably Galileo who should be credited with the realization. It captures that strange experience you may have had when you're in a train alongside another train at a station and then your train starts moving. Until the platform is revealed, it's impossible to tell which train is moving.

When this is applied to our clocks, it leads to a rather strange outcome. As far as I am concerned in my spaceship, it is the Earth that is whizzing by me at 3/5 the speed of light. The same analysis that I performed above implies that I will calculate that it is your clock on Earth that is going slower, not my clock. Time, it turns out, is far less obvious a concept than we experience in everyday life.

The whole thing seems so bizarre as to be unbelievable. How can the spaceship clock be ticking slower than the Earth clock and at the same time the Earth clock be ticking slower than the spaceship clock? Yet if I accept the indisputable observation that the speed of light is constant regardless of how I measure it, the mathematics leads me to this conclusion. It's one of the reasons I love mathematics. It is like a rabbit hole of logic dropping you into unexpected wonderlands.

It's not just the ticking clock on the spaceship that appears to be slowing down from the perspective of Earth. Anything that keeps track of time must also slow down. If I am sitting on the spaceship I cannot tell that my clock is doing anything strange. So it means that anything that measures time on the spaceship will be similarly affected, including the quartz pulsating in my wristwatch, the Prokofiev on the spaceship's radio, the aging of my body, my brain's neural activity. On board the spaceship, I will not be aware that anything strange is happening because everything on board the spaceship will be ticking at the same rate.

But from your perspective on the Earth, it appears that my watch is losing time, the Prokofiev has turned into a deep-sounding dirge, I am aging more slowly, and my neurons aren't firing quite as fast as they usually do. Time and the sensation of its passing are relative. If everything is slowing down or speeding up at the same rate, I can't tell the difference. Everything seems normal on board my spaceship. The curious thing is that when I look down at you, I see everything around you slowing to a snail's pace.

SPEED UP AND LIVE LONGER

A rather striking example of this relative difference in the passage of time is the strange case of muon decay that I encountered in the Second Edge. When cosmic rays hit the upper atmosphere, the collisions create a shower of fundamental particles, including the muon, a heavy version of the electron. These muons are not stable and very quickly decay into more stable forms of matter.

Scientists talk about something called half-life. This is the amount of time it takes for the population of muons to be reduced by half because of decay. (Knowing when the decay will happen for a particular particle is still something of a mystery.) In the case of muons, after 2.2 microseconds, half of these particles on average will have decayed away.

The speed at which they decay should mean that, given the distance they must travel to the surface of the Earth, not many will survive the journey. But scientists detected far more muons than they expected. It turns out that these particles are traveling at close to the speed of light. So the half-life of the muon is actually longer than we'd expect when measured by a clock on Earth. The muon's internal clock is going slower than the clock on Earth, and hence, since not so much time has passed in the frame of reference of the muon, the 2.2 microseconds that it takes for half the muons to decay takes much longer than 2.2 microseconds measured by a clock on the Earth's surface.

How does this work from the perspective of the muon? The clock on board the muon is running normally, and it's the clock on Earth that is running slowly. So, from their perspective, what causes more muons to reach the surface of the Earth than expected? The point is that it isn't only time but also our perception of distances in space that is altered by looking out the window of a spaceship traveling close to the speed of light. Distances shrink when you are moving, so the distance from the outer atmosphere to the surface of the Earth, from the muons' perspective, is much shorter than it is from our own perspective. So the muon doesn't think it has as far to go, and more reach their destination.

Is this a strategy I can use to add a few more days to my finite life? Can I cheat my own half-life? The trouble is that, as I explained, everything slows down on my speeding spaceship. I'm not going to squeeze any more time out of the universe to solve the mathematical problems I'm working on by speeding along near the speed of light, because while my body may age more slowly, my neurons will fire more slowly too. The principle of relativity means that as far as I am concerned, I appear to be at rest and it's everything else that is speeding by.

RELATIVITY DOGS

What does it mean for two events to happen at the same time? This was a problem Einstein confronted when working on patents to synchronize time. It turns out that the answer will depend on your frame of reference.

Let's start with a scene from a fictional movie called *Relativity Dogs,* in homage to Quentin Tarantino. It takes place on a train (as many things do in relativity). Two people with identical guns are standing at either end of the train. Exactly halfway between them is a third member of the gang. The train is racing through a station. A police officer is watching the scene. Let me first consider the situation on the train. As far as the gang members are concerned, the train can be considered at rest. The guns go off. The bullets hit the man in the middle at the same time. The speed of the bullets and the distance they have to cover is the same, and as far as everyone on the train is concerned the gunmen both shot at the same moment. Indeed, the

victim saw light flash from the guns at the same moment, just before being hit by the bullets.

But what about the perspective of the police officer? Let's suppose the victim passes the police officer at precisely the moment both flashes of light reach the victim, so that the police officer witnesses the flashes at the same time too. But then the officer begins to wonder. Although they are the same distance away from him now, when the guns went off, the one at the front of the train was actually nearer to him. So the light had a shorter distance to travel than the light of the gun at the back of the train. In which case, since the speed of light is constant, if the light arrived at the same time it must have left the gun at the back of the train earlier than it left the gun at the front. So to the police officer it seems that the gunman at the back of the train shot first. But if I put another officer on a train going in the opposite direction, then everything is reversed and the second police officer will conclude that the gun at the front of the train must have gone off first.

So who shot first? The gunman at the back shot first from the perspective of the policeman on the platform; but the gunman at the front shot first from the perspective of the policeman on the train speeding in the opposite direction. To talk about which gun was fired first is therefore meaningless in an absolute sense. Time takes on different meanings for different frames of reference. It turns out there is something that is absolute for all of the observers, but it requires the combination of time and space.

The trouble is that I have been trying to measure the distance between two objects, and this changes according to my own movement relative to the two points. The time between two events also changes. But if I can identify a new distance that measures the distance in time and space, I can get something invariant—that is, not dependent on who is doing the measuring. This was the great idea of mathematician Hermann Minkowski, Einstein's former teacher at the Polytechnic in Zürich. On hearing about Einstein's ideas, he immediately understood that the high-dimensional geometries discovered by German mathematician Bernhard Riemann fifty years earlier were the perfect stage for Einstein's theory.

For those happy to contemplate a formula, the distance between an event happening at location (x_1, y_1, z_1) at time t_1 and an event happening at location (x_2, y_2, z_2) at time t_2 is defined by

$$\sqrt{(x_1-x_2)^2 + (y_1-y_2)^2 + (z_1-z_2)^2 - c^2(t_1-t_2)^2}$$

The first three bits of the equation

$$(x_1-x_2)^2 + (y_1-y_2)^2 + (z_1-z_2)^2$$

make up the usual distance measured in space (using Pythagoras's theorem). The last bit is the usual measurement of the difference in time. Your first intuition might be to add these two distances together. The clever thing Minkowski did was to take the second away from the first. This creates a very different sort of measure, leading to a geometry that doesn't satisfy the usual laws of geometry as developed by the Greeks. It views the universe not as three-dimensional space animated in time but rather as a four-dimensional block of something called space-time, whose points are located by four coordinates (x, y, z, t)—three for space and one for time. Minkowski introduced this new way of looking at the universe two years after Einstein's announcement of his special theory of relativity.

If the formula leaves you none the wiser, don't despair. Einstein too was rather suspicious of what he regarded as something of a mathematical trick. But Minkowski's four-dimensional geometry would provide a new map of the universe. As Minkowski declared, "Henceforth space by itself, and time by itself, are doomed to fade away into mere shadows, and only a kind of union of the two will preserve an independent reality."

This was Einstein's response to the mathematization of his ideas: "Since the mathematicians pounced on the relativity theory I no longer understand it myself." But he soon realized that mathematics was the best language for navigating this strange new universe called space-time.

The power of measuring distance in space-time is that, regardless of your frame of reference, everyone agrees on the measurement. If you measure time and distance separately you will get different values depending on how you are moving relative to the things you are measuring. But combine space and time and now your movements don't matter. So Newton was right in some respects: you do need a background that is absolute. His mistake was to consider time and space separately. Post-Einstein, we must consider the two simultaneously. And it's this mixing of the nature of time and space that makes the question of what happened before the Big Bang really interesting.

For a start, it forces us to consider time in a very different light. We should regard the universe as a block of space and time in which the ideas of "before" and "after" are as questionable as pinpointing which point of space is in front of another—it depends on your perspective. This is very disconcerting. For both policemen in my fictional movie scene, there is a moment when one of the gunmen still hasn't squeezed the trigger. Perhaps there is time for him to stop and consider his actions, to decide not to shoot and let the other person be responsible. But hold on! For the policeman on the platform that decision is in the hands of the man at the front of the train. But for the policeman on the train going in the opposite direction, it's the gunman at the back. Does that actually mean the future is not really in our hands at all?

Usually when I draw a graph of distance against time, the time axis runs along the horizontal and the distance, traveled by a ball, say, is represented on the vertical. But space-time doesn't allow me to divide time and space up so neatly. When I think of space-time as a block, I must be careful not to think of one privileged direction in this block representing time and three other independent directions keeping track of space. Two different directions in this space-time can represent the time dimension. It just depends on how you are moving through space. Time and space are mixed in this new view.

This is a real challenge to my intuition about the universe. Suppose I am speeding away from you on a spaceship. If I draw a line connecting all simultaneous events from my perspective in this space-time geometry, they will be a very different set of lines from yours.

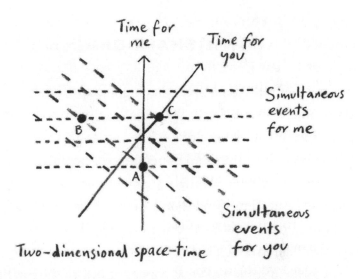

Two-dimensional space-time

On your timeline, events A and B happen simultaneously and C happens later. However, for me, B and C are simultaneous events, while A happened earlier. If A and C are causally linked then for everyone's timeline A will always happen before C. But B is not causally connected to either A or C, so there are timelines that place event B before A, or alternatively after C.

In Hindi and Urdu the word *kal* is used to mean both yesterday and tomorrow. In the quote from *Midnight's Children* at the head of this chapter, Salman Rushdie joked that people who have the same word for yesterday and tomorrow cannot be said to have a firm grip on time. But could they be on to something? The idea of "before" or "after" is not as clear as some languages imply.

And yet, even with this mixing of space and time, time does have a different quality from space. Information cannot travel faster than the speed of light. Causality means that we can't place ourselves at a point in space-time where the bullet hits the victim before it was fired. There are constraints on how these timelines can be drawn. My intuition about space and time is not going to help me. Instead, as Einstein reluctantly conceded, I have to rely on mathematics to lead me.

THE SHAPE OF TIME

Just as I can talk about space having a shape, I can also talk about space-time having a shape. My initial instinct is to think of time as a line, which makes it very difficult for me to think of anything other than a line that is finite and therefore has a beginning and an end, or a line that is infinite. But there are other possibilities. Since time and space make up four dimensions, I must consider shapes that I can't see. I need mathematics to describe them. But I can picture shapes representing parts of space-time that enable me to understand what I mean by the question of what happened before the Big Bang. For example, imagine there was only one spatial dimension, so that space-time is two-dimensional. This creates a surface that I can see, like a rubber sheet that I can play with and wrap up in interesting ways.

I think most people's model of two-dimensional space-time would be an infinite flat sheet with time extending infinitely backward and forward and space also one-dimensional and infinite. But, as I explored in the last Edge, space could be finite. I could wrap space up to make a circle and then time would extend this circle to make space-time look like a cylinder. I could, of course, join the cylinder up to make a bagel- or torus-shaped space-time. That would make time finite too. In this model of space-time I could loop around and return to a previous moment in history. Logician Kurt Gödel proposed solutions to Einstein's equations of general relativity that have this feature. Gödel, as I shall explain in my final Edge, greatly enjoyed messing with people's logical expectations. But Gödel's circles of space-time are generally regarded as curiosities, because going back in time causes too many problems when it comes to causality.

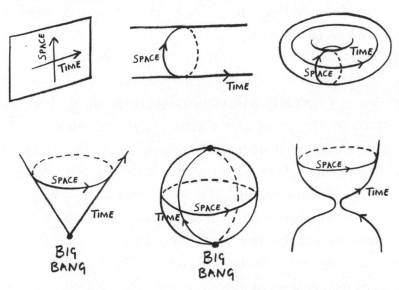

Possible shapes for a two-dimensional space-time

To get a more realistic picture of space-time, I need to create a geometry that takes into account our current model of the history of the universe, which includes a beginning: the Big Bang. To see this point in my two-dimensional space-time universe, I could wrap up the surface so that it looks like a cone. The universe, which is just a circle, shrinks as I go back in time until I hit the point at the end of the cone. This is where time begins. There is nothing before that. No space. No time. Just a point of infinite density. This is a good model for something rather like the Big Bang.

Or, rather than pinching space-time at a point, perhaps it could look like a sphere—like the surface of the Earth. This has its own implications for how I answer the question of what happened before the

Big Bang. If I head south along a line of longitude, then when I hit the South Pole there is a sudden flip and I find myself on another line of longitude on the other side. But these are just numbers that we give to these points on the Earth's surface. It doesn't actually represent a discontinuous flip in the shape, just a jump in the way that we measure it.

So by changing the coordinates, what appears to be a singular point can look quite smooth. This is one of Hawking's ideas for time. Perhaps I should try to embed space-time in a shape such that this point that seems to show time stopping is actually just the south pole of the shape. After all, how do you answer the question, what's south of the South Pole? The question doesn't really make sense.

It is striking that often when a question arises to which it seems we cannot know the answer, it turns out you need to acknowledge that the question is not well posed. Heisenberg's uncertainty principle is not really an expression of the fact that we cannot simultaneously know both the momentum and position of a particle, so much as that these two things don't actually exist concurrently. Similarly, attempts have been made to show that the question "What happened before the Big Bang?" isn't one whose answer cannot be known. Rather, the question does not really make sense. To talk about "before" is to suppose the existence of time, but what if time existed only after the Big Bang?

By picturing the shape of space-time I begin to get a sense of why many scientists have dismissed the question of understanding time before the Big Bang as meaningless. But there are other shapes that allow time to have a history before the Big Bang. What if the cone, instead of coming to a point and ending, actually bounced out of a contracting universe before the Big Bang? To truly get a sense of time's history as I head back toward the Big Bang, I need to understand what happens to time as it approaches a point with increased gravity. This was Einstein's second great discovery: that gravity also has an effect on the ticking clock of time.

WHY SKYSCRAPERS
AREN'T GOOD FOR YOU

Einstein's second assault on the nature of time came when he threw gravity into the mix. His general theory of relativity, developed between 1907 and 1915, describes the very geometric nature of gravity. Gravity is not really a force but a property of the way this four-dimensional sheet of space-time is curved. The moon orbits the Earth because the mass of the Earth distorts the shape of space-time in such a way that the moon simply rolls around the curved shape of space-time in this region. The force of gravity is an illusion. There is no force. Objects are just free-falling through the geometry of space-time, and what we observe is the curvature of this space. But if massive bodies can distort the shape of space, they can also have an effect on time.

This was yet another one of Einstein's great revelations, and it is based, once again, on a principle of equivalence. The strange consequences of special relativity are teased out of the principle of relativity, which states that it is impossible to tell whether I am moving or my environment is moving past me. Einstein applied a similar principle of equivalence to gravity and acceleration.

If you are floating out in space, in a spaceship with no windows, and I place a large massive planet underneath the spaceship, you will be pulled toward the floor. This is the force of gravity. But if instead I accelerate the ship upward, you will experience exactly the same sensation of being pulled toward the floor. Einstein hypothesized that there is no way to distinguish between the two: gravity and acceleration produce the same effects.

Let's suppose that our spaceship is as tall as the Shard in London. I'm going to place a photon clock at the bottom of the spaceship and another at the top. Next to each clock I'm going to station an astronaut who will help me compare the running of these two clocks.

The astronaut at the bottom of the spaceship is going to send a pulse of light to the astronaut at the top each time his clock ticks. The astronaut at the top can then compare the arrival of these pulses with the ticking of her clock. Without acceleration or gravity, the arrival of the pulses and the ticking of the clocks will be in sync. However, let me now accelerate the spaceship in the direction of the top of the spaceship. A pulse is emitted at the bottom of the spaceship, and, as the spaceship accelerates away, the light has farther to travel each time, so that it takes longer and longer for each pulse to reach the top of the spaceship, and the astronaut at the top will receive the pulses at a slower rate. This is similar to the Doppler effect we experience with sound, where moving away from the source causes the frequency to decrease, resulting in a lower pitch. But in this case it is important to note that the spaceship is accelerating rather than just flying along at a constant speed.

The clock at the bottom of the spaceship will appear to be going slower than the clock at the top. But what if I reverse the experiment and get the astronaut at the top of the spaceship to send pulses down to the bottom of the spaceship? Because the astronaut at the bottom is accelerating toward the pulses, he is going to receive them at a faster rate than the pulses of light he is sending. So he will confirm that his clock is running slower than the clock at the top of the spaceship.

The interesting conclusion of the experiment comes when I re-place acceleration with gravity. According to Einstein's principle of equivalence, whatever effect acceleration had on the clocks in the spaceship, the effect of gravity must be the same. So when I place a large planet at the foot of our Shard-sized spaceship, the impact is the same as if the spaceship were accelerating through space: clocks run slower at the foot of the Shard than they do at the top.

Since the aging body is a clock, this means that you age slower the closer you are to the center of the Earth—people working at the top of the Shard in London are aging faster than those on the ground floor.

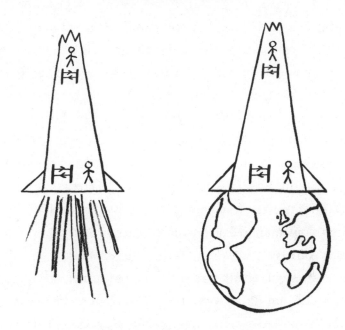

Acceleration and gravity have the same effect:
slowing the clock at the bottom of the Shard spaceship.

Of course, the difference in the speed of the clocks is extremely small at this scale, but it makes a significant difference if we compare the ticking of atomic clocks on the surface of the Earth with those in orbit on satellites. The difference in gravity experienced by the two clocks results in them ticking at different rates. Since these atomic clocks are integral to the functioning of GPS, it is essential that the effect of gravity on time is taken into account if the systems are to be accurate.

ASYMMETRICAL TWINS

There is a classic story that reveals the strange nature of time in Einstein's new world. It involves sending twins, or actually one twin, on a space journey. It is particularly close to my heart because I have identical twin girls, Magaly and Ina. If I send Ina off on a spaceship traveling at close to the speed of light and then bring her back to Earth, the physics of relativity implies that, although she will think that she's been away for only ten years, her twin sister's clock on Earth has raced ahead, so that Magaly is now in her eighties.

To truly understand the asymmetric nature of the story, I need to take into account Einstein's revelation about the effect of gravity and acceleration on time. Once Ina is traveling at a constant speed close to the speed of light, Einstein's first revelation declares that neither twin can tell who is moving and who is still. Ina will think Magaly's clock is running slower, and Magaly will think Ina's is slower. So why does Ina return younger?

Ina returns younger because she has to accelerate to get to her constant speed. Similarly, when she turns around she needs to decelerate and then accelerate in the opposite direction. This causes her clock to slow down relative to that of her twin on Earth, who doesn't accelerate. This asymmetry results in Ina heading into Magaly's future. If I sent both twins off in spaceships in opposite directions and brought them back together, then they'd be the same age—and everyone on Earth would have aged quicker.

Einstein's general theory of relativity revealed that time as well as space gets pulled about by things with mass. Gravity is actually the distortion of this space-time surface. If something has mass, it curves the surface. The classic way to imagine this is to consider space-time as a two-dimensional surface, and the effect of mass as

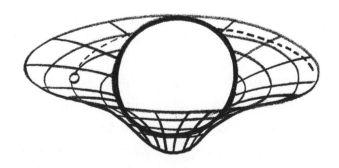

that of placing a ball on this surface. The ball pulls the surface down, creating a well. Gravity can be thought of as the way things get pulled down into this well.

This distortion of space-time has an interesting effect on light. Light follows the shortest path between two points—the definition of a "straight" line. But I am now talking about lines in space-time, where distance is measured using Minkowski's formula that includes the coordinates of space and time. Weirdly, using Minkowski's formula, it turns out that the distance between two points in space-time is reduced if the light takes longer to get there.

So in order to find the shortest space-time path, light will follow a trajectory that tries to balance a minimizing of distance traveled against a maximizing of time taken. By following a trajectory such that the light particle is essentially free-falling, the clocks on board the particle will go faster. Pull against gravity and you are accelerating and thus slowing your clock down. So Einstein's theory predicted that light would be bent by the presence of a large mass. It was a highly unexpected prediction, but one that could be tested—a perfect scenario for a scientific theory.

British astronomer Arthur Eddington's observations of light from distant stars recorded during the solar eclipse of 1919 provided convincing evidence for this picture of a curved space-time. The theory predicted that the light from distant stars would be bent by the gravitational effect of the sun. Eddington needed the eclipse to block out the glare of the sun so that he could see the stars in the sky. The fact that the light did indeed seem to bend round objects of large mass confirmed that the shortest paths weren't Euclidean straight lines but curved.

We experience the same effect on the surface of the Earth. An airplane flying from London to New York takes a curved path passing over Greenland rather than the straight line we'd expect by looking at a map of the Earth. This curved line is the shortest path between the two points on the Earth. Light too found the shortest path from the star to Eddington's telescope on Earth.

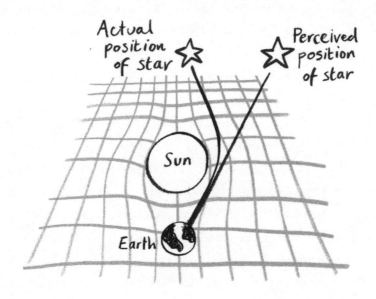

Eddington announced his experimental evidence confirming Einstein's ideas on November 6, 1919. Within days, newspaper headlines across the world trumpeted the great achievement: "EINSTEIN THEORY TRIUMPHS: Stars Not Where They Seemed to Be, but Nobody Need Worry" announced the *New York Times*; "Revolution in Science" declared the *London Times*. This was probably the first time in history that a scientific achievement was given such public exposure. Heralded by journalists as the new Newton, the obscure forty-year-old Einstein shot to international fame.

If you are finding all this warping of space and time bending your brain, don't despair. You are in good company. After Eddington announced his discovery that light bends, a colleague came up to congratulate him: "You must be one of only three people in the world to understand Einstein's theory." When Eddington failed to respond, the colleague prompted, "Come, come, there is no need to be modest."

"On the contrary," replied Eddington, "I was trying to think who was the third."

Trying to say what happens to time as we head back toward the Big Bang would test even Eddington's understanding of Einstein's theory.

10

There is a place where time stands still. Raindrops
hang motionless in air. Pendulums of clocks float
mid-swing. Dogs raise their muzzles in silent howls.
Pedestrians are frozen on the dusty streets, their
legs cocked as if held by strings. The aromas of
dates, mangoes, coriander, cumin are suspended
in the space.

—Alan Lightman, *Einstein's Dreams*

'M RATHER FOND of the design of my watch. The face is a simple
brown square sitting inside a circular silver case. I like the symmetry of the shape, and the subtle tension between the square and the circle. It's not expensive, which is a good thing, as I tend to lose watches.

My last watch slipped off my wrist when I was kayaking across a lake next to the glacier on Mount Cook. I tried to rescue it but the water was so bitterly cold that I couldn't leave my hand in for more than a couple of seconds. (Not that I had anything to measure time with anymore.) The watch had disappeared below the surface and its mechanism is by now presumably rusted and frozen, stopped by the glacial waters flowing off Mount Cook.

If I want an environment to freeze the ticking of my current watch, dropping it into the extreme regions of space discovered within Einstein's equations for general relativity would be a good bet. Only I'd need a spaceship, not a kayak, to explore the strange phenomena known as black holes.

One reason why the universe is so fascinating is that thanks to gravity, atoms attract other atoms. And so unless everything is in a perfect balance, we see a movement of matter toward other bits of matter. But the interesting thing about gravity is that its pull grows stronger as things get closer. This attraction is what leads to the creation of stars like our sun, but also to the potential for more catastrophic events.

The simplest atom is hydrogen, made up of one electron and one proton held together by the electromagnetic force. If I have two atoms of hydrogen, the force of gravity will pull them together. And as the atoms are pulled closer and closer they start to collide. Increase the number of hydrogen atoms and the collisions become more and more energetic until the atoms no longer simply bounce off each other; eventually, the conditions for nuclear fusion are satisfied and I get a star. Hydrogen atoms fuse to create helium atoms. In the process, energy is given off, creating outward pressure as the energy is dissipated. This is the energy that we depend on for life on Earth. Stars remain stable because the inward gravitational pull is counterbalanced by the outward pressure of the energy released in the process of fusion.

At some point all the hydrogen gets used up. Some stars will continue to fuse helium into atoms further up the periodic table. Many of the atoms we find on Earth, such as iron, oxygen, or even the carbon that is crucial for making life, are all products of this continuing fusion of lighter atoms. Eventually, the star isn't able to sustain this fusion—the fuel is spent. Gravity takes over once again, and as the star contracts, quantum physics comes into play. If I confine particles in a smaller and smaller space then I know a lot about their position. According to Heisenberg's uncertainty principle, this must be balanced by huge uncertainty about the velocity of the particles. The movement of the particles away from each other results in a second period of stability that counteracts the pull of gravity, resulting in what is known as a white dwarf.

In 1930, the Indian physicist Subrahmanyan Chandrasekhar challenged this theory. Stuck on a boat sailing from India to England where he was heading to do his doctoral studies in Cambridge, Chandrasekhar suddenly realized that special relativity puts a speed limit on how fast particles can move. So if the mass of the star is great enough, gravity will win out over the speed limit and the star will continue to collapse, creating a region of space of increasingly high density. His calculations, made on board the ship, revealed that any star more than 1.4 times the mass of our sun would suffer such a fate. The supernovae responsible for creating elements like gold and uranium are the result of these cataclysmic collapses.

In 1916, German physicist Karl Schwarzschild first deduced from Einstein's equations that space around these points of high density would be severely warped, so much so that light would be trapped inside. Einstein didn't believe such points could exist, but now Chandrasekhar's work revealed that such points are not just theoretically possible but would be the result of collapsing stars. One way to get a sense of why light might get trapped is to consider what happens when you throw a ball in the air. On the surface of the Earth it is possible to launch a ball fast enough to escape the gravitational pull of the Earth. The speed at which you need to throw the ball is called its escape velocity. But imagine gradually increasing the mass of the Earth. The speed required to escape its gravitational pull would also increase until, at some point, the mass of the Earth would be so enormous that the ball would need to be launched at a speed faster than the speed of light in order to escape. At this point the ball is trapped. It cannot go beyond a certain point before it is pulled back to Earth.

This scenario follows our classical pre-Einstein view of gravity. Indeed, there is evidence that the idea of light being unable to escape the effects of gravity had been predicted before Einstein ever appeared on the scene. Toward the end of the eighteenth century, Pierre-Simon Laplace and John Michell had already flirted with the proposal that light could be trapped by massive objects. A hundred years later, Michelson and Morley discovered that light always travels at the same speed in a vacuum. This meant that light would not actually behave like our ball. Gravity won't slow it down, as Laplace and Michell thought.

Einstein's conceptions then complicated things once again. Although gravity might not slow light down, in Einstein's theory massive objects could warp space so much that light could not find a path to escape the region surrounding the object. The light in effect gets bent back on itself. In 1967, American physicist John Wheeler whimsically referred to these regions as "black holes." Richard Feynman thought the name was obscene; in French, *trou noir* has other connotations. But the name stuck.

As you move away from the center of the collapsed star, the effect of gravity decreases. This results in a boundary, shaped like a sphere, with a black hole at its center that represents a point of no return—any light or anything inside this boundary will be trapped because it doesn't have enough speed to escape. This sphere is known as the event horizon of the black hole, as anyone outside the sphere cannot witness the events occurring on the inside.

Black holes have been a subject of hot debate since their existence was first proposed following Einstein's publication of his equations for gravity in 1915. Some believed that stars, as they collapsed, would somehow avoid such no-go regions. Perhaps the star would throw off mass. That is certainly possible, but a star that is twenty times the mass of the sun would need to throw off ninety-five percent of its mass to avoid becoming a black hole, which seems unlikely. Still, many scientists insisted that these bizarre and unfathomable regions of space-time would not occur.

In 1964, the first potential example of such a high-density region was identified in the constellation of Cygnus. Astrophysicists called it Cygnus X-1. By 1971, calculations of its mass and concentration led to the conjecture that it was a black hole. Not everyone was convinced.

In fact, one notable person made a bet in 1975 to the effect that Cygnus X-1 was not a black hole: Stephen Hawking. This was somewhat odd, given that he'd dedicated much of his research to probing the nature of black holes. If Cygnus X-1 turned out to be the first example of a black hole, all Hawking's theoretical musings would have been justified.

As he explained in *A Brief History of Time*, the bet was an insurance policy. Betting against your football team winning the final of the FA cup is a win–win situation: if your team loses, at least you benefit financially. If it turned out that his life's work on black holes had been fruitless, then at least he'd won the bet. His prize? A subscription to *Private Eye* magazine so he could distract himself from the misery of his failed research. He made the bet with fellow cosmologist Kip Thorne. Convincing evidence that Cygnus X-1 was indeed a black hole would in turn win Thorne a subscription to the journal of his choice. He opted for *Penthouse* magazine.

By 1990, evidence was finally found that Cygnus X-1 was indeed a black hole. It is estimated to have a mass of 14.8 times the mass of the sun and to be too compact to be a star and is believed to have an event horizon of forty-four kilometers. Inside this spherical ball, whose diameter stretches the distance between Oxford and Cambridge, no light can escape. Given all the data, Hawking conceded the bet. Thorne got his subscription to *Penthouse*, much to his wife's chagrin.

There is something mathematically disturbing about these black holes. As the star collapses, creating a point of high density, nothing seems to counter the continuing pull of gravity. So why would it not continue collapsing, becoming smaller and smaller and more and more dense with nothing to stop the implosion? Could it keep collapsing until it created a singular point of infinite density?

Einstein himself tried to prove the absurdity of this conclusion. Eddington could see what the math was implying but balked at the implications. "When we prove a result without understanding it—

when it drops unforeseen out of a maze of mathematical formulae—
we have no grounds for hoping that it will apply," he said. It wouldn't
be until 1964 that Oxford mathematician Roger Penrose finally proved
that such singular points were a necessary consequence of the theory
of general relativity.

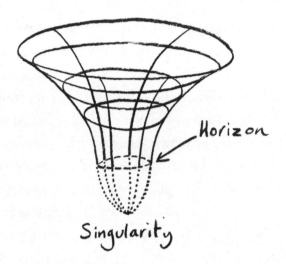

A black hole in two-dimensional space-time. The event
horizon is a circle inside of which we cannot penetrate.

In collaboration with a young Stephen Hawking, Penrose went on
to prove that the same infinite density is predicted when we rewind
the universe back to the Big Bang. Both black holes and the Big Bang
are examples of a mathematical entity called a singularity. Singularities encompass a whole range of situations where it is impossible to
work out what's happening. A singularity is a point at which our ability to model the scenario breaks down—a place where we have to
throw up our hands and declare that we do not know.

SINGULARITIES

Singularities are points at which mathematical functions crash. A function in mathematics is a little like a computer program. You input numbers, the function calculates away, and then it spits out an answer. Mathematicians often represent functions visually by using graphs. The input is given by a number on the horizontal line, we mark the output above the horizontal line, and the result is a curve.

Consider the function that inputs the distance from a massive object and outputs the gravitational pull experienced at that point due to the massive object. Newton realized that the greater the distance from the object, the weaker the pull. He discovered a very precise relationship between the pull and the distance. If I am a distance x from the planet, Newton's function calculates that the pull of gravity is proportional to $1/x^2$. This is what we mean by an inverse square law. I can draw a picture of this function.

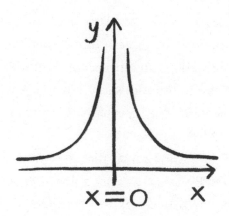

Graph of $1/x^2$. The function has a singularity at $x = 0$.

But something interesting happens as I get closer to the object. The force increases until, when I am at distance $x = 0$, the output is infinite and the graph doesn't have a value. Of course, in reality I am measuring the distance from the center of gravity of the planet, and once I reach the surface of the planet the function and graph change because, as I pass through the surface, parts of the planet start pulling in the other direction. When I hit the center of gravity, all the pulls balance out and I experience zero gravitational pull. But what if I replace the planet with a black hole, a region in space where all the mass is meant to be concentrated at a single point? Now this point has infinite density, and I can approach it, and as I hit the point the gravitational pull is infinite.

The fact that the function doesn't really make sense at $x = 0$ is what mathematicians call a singularity. There are different sorts of singularities, but they all have a point at which the function doesn't give a sensible output, or where there is a sudden discontinuous jump from one value to another.

A very homespun example of a singularity can be realized by taking a coin and spinning it on a table. If there were no friction on the table or air resistance, the coin would spin forever at a constant speed. However, because energy is dissipated, the coin does not spin forever. Instead, the angle of the coin to the table decreases, and, intriguingly, the speed of the rotation increases proportionally. As the angle approaches zero, the speed eventually becomes infinite. The final stages of the spinning result in the coin shuddering as it falls and there is a whirring sound whose frequency rapidly increases until suddenly the coin comes to a juddering halt.

The equations of motion reveal that the speed of the coin's rotation increases at such a rate that after a finite amount of time it hits infinite speed. This is what we hear when the frequency of the sound increases. The spinning coin is an example of a singularity. Of course, other effects come into play to stop the full realization of this mathematical infinity, but it does reveal that you don't have to disappear down a black hole for physics to produce infinities.

Singularities are typically moments beyond which it is impossible to make predictions. It isn't just in physics that these singularities can emerge. There is a famous example, published in 1960 by Heinz von Foerster, Patricia Mora, and Lawrence Amiot, that predicts a serious singularity here on Earth. The rate of population growth, following the pattern observed up to 1960, indicated that the population of the planet would become infinite on November 13, 2026. A Friday, it so happens, if you're at all superstitious.

The simplest model of population growth posits that it is exponential. For example, a species might double in number every fifty years. With such a model, the population quickly explodes but never becomes infinite. But the authors' analysis of previous data revealed that the period of time it took for the human population to double was getting shorter and shorter.

The population took 1,650 years to double from 250 million in AD 0 to 500 million in 1650. Two hundred years later, it reached a billion. The next doubling took only eighty years. The population hit four billion in 1976. The rate of growth is faster than exponential. So in 1960, with the data to hand, the authors estimated that the population of the Earth is due to hit a singularity a decade or so from now.

Another example of this super-exponential growth is the rate of increase in computer power. There is a dictum called Moore's law, which states that computers double in power every eighteen months. With such an increase, computers are powerful but never hit a singularity. But others have suggested that just as the time it takes for the population to double seems to be getting shorter, the same applies to technology. The possibility of a technological singularity has given rise to something called the Singularity movement. Popularized by inventor and futurist Ray Kurzweil in his book *The Singularity Is Near*, the singularity is due to hit humanity in 2045. At this point, Kurzweil believes, humans will be able to create artificial intelligence that exceeds our own. This moment will be accompanied by a breakdown in our ability to predict what life after such a singularity will be like.

We have to be careful about mathematical equations, because there may be a hidden piece that becomes significant only when we approach the singularity, and which will then play a large role in preventing any physical realization of this infinity. This clearly occurs in the case of population growth. The finite area of the Earth's surface will, beyond some threshold, constrain how big a population can be produced. At this point we will be more likely to see the chaotic population dynamics that we explored in the First Edge start to dominate. But with any luck we will constrain ourselves before then.

Similar factors could come into play with the Big Bang and black holes. Some have suggested that the equations of general relativity don't apply in these extreme settings. It is possible that we may need to introduce another term into Einstein's equations for gravity that would affect the equations only when we approach the singularity. It is like the subtle change that Einstein needed to introduce when considering things traveling near the speed of light. As I explained in the previous chapter, the formula for the speed of a man running on a moving train relative to the platform is given by adding together the speed of the man and the train, but then you need to divide by a second formula. At low speeds, this second formula is so close to 1 that it has little effect, which is why pre-Einstein scientists believed that the speeds were simply added together. But close to the speed of light, a different behavior dominates. The same could be true close to the Big Bang or a black hole.

But if the universe does contain these singular points of infinite density, what effect do they have on time? Einstein discovered that increased gravity slows time down. So what would happen to my watch if I were to travel to the center of a black hole?

WHAT'S INSIDE A BLACK HOLE

If I throw my watch into a black hole, something strange occurs. If I stay on Earth and observe my watch being sucked into the black hole, there is a moment when time seems to stop. My watch goes slower and slower until it doesn't tick anymore. Eventually the image of my watch is frozen, and then fades. The event horizon surrounding a black hole is like a bubble in space beyond which time seems to break down. From my perspective outside the black hole, there is no "after." Could this be the reverse of what happens to time as we head back to the Big Bang? In that scenario would time not have a "before"?

But remember, this is from my perspective, as I compare time on Earth to the time I can see on my watch as it heads toward the black hole. How do things look if I wear my watch on my wrist and join it on its trip? My experience is now very different. From Earth, when my watch hit the event horizon of the black hole it appeared to stop ticking. But from my new perspective, as I cross the event horizon my watch seems to continue ticking. Indeed, I won't actually be able to detect when I have crossed this line of no return.

That's not to say that all is well for me as I hurtle through space. The infinitely dense center is not so easily ironed out. As I head feet first toward the center of the black hole, gravity pulls harder on my feet than my head, stretching me out until I am spaghettified. Then, a short while later, everything will be crushed, including my watch, and time will come to an end. Like a line drawn across a page, time reaches the edge and there is nowhere for it to go, while elsewhere it seems to tick on obliviously.

I guess we all have a similar fate awaiting us. When we die, time stops for us, but we know that time will continue on unawares. Just as I cannot experience my own death, I wouldn't experience arriving at a singularity in space-time.

Just in case I do slip into a black hole, a physicist friend offered me some advice: as with quicksand, the best thing is not to struggle. If I free-fall toward my destiny, I will survive longer. This sounded very counterintuitive. My friend told me to think of what happened to my watch under the effects of gravity and acceleration. If I struggle, and try to pull away, I'll find time running slower where I am. Increased acceleration, like gravity, slows clocks down. But that won't be good. Remember my accelerating daughter Ina, who raced into the future while her twin Magaly stayed behind on Earth? By accelerating her spaceship, Ina only experienced ten years while Magaly was in her eighties when Ina returned to Earth. By not struggling or accelerating, Magaly had experienced many more decades of life than her space-traveling sister. A similar push and pull is at work here. So if I struggle when I arrive at the singularity I will have aged less. But that's not much good: I'll be extinguished after having experienced less of life. So it's better not to struggle and to enjoy my last few moments of free falling.

The interesting thing is that someone watching from Earth will never know what is going on inside the black hole. From their perspective, time appears to stop as soon as I cross the event horizon; and for them there is no telling what comes after, although there is an after for me as I hurtle on toward spaghettification. Asking what happens next appears to be a sensible question. It has an answer. The only trouble is that if you are outside the event horizon, the laws of physics deny your access to that answer.

The same principle was thought to apply to the challenge of knowing what, if anything, happened before the Big Bang. Time comes to an end inside a black hole. But the Big Bang is like the collapse of a star into a black hole played in reverse. So does time also come to an end as I head back toward the Big Bang? In other words, does time have a beginning—and does that time itself start at the Big Bang?

These black holes offer a serious challenge to our concept of time.

They also appear to contradict one of the other discoveries of modern physics: that information is never lost.

THE ULTIMATE PAPER SHREDDER

A rather remarkable consequence of the laws of quantum physics is that they are reversible. This means that information is never lost. It is a very counterintuitive proposition. If I took a subscription to *Private Eye* and a subscription to *Penthouse* magazine and I burned a year's worth of both, then it seems highly unlikely that I could ever determine which pile of ashes corresponds to which magazine. But provided I have complete information about all the atoms and photons involved in my bonfire, it is theoretically possible to rewind the process and recover the information contained inside the magazines. It would, of course, be extremely difficult in practice, but the science asserts that there is nothing irreversible about the process of burning. The laws of physics work both ways.

The existence of black holes challenges this idea. If I throw one magazine into one black hole and the other into another black hole, it will be impossible to find out which magazine went into which black hole. Black holes seem to be the ultimate paper shredders—information appears to be genuinely lost. But is it really?

Black holes are particularly tantalizing in my search for things we cannot know, because once something has disappeared behind the event horizon—the boundary beyond which light cannot escape—it appears that I actually lose all information about it. If I take the casino die that I've been carrying with me and I toss it into the black hole, I can never know how it will land. There could be a desk on the other side on which it lands and comes up with a 6. Someone on the other side could see this but never be able to communicate the result.

According to general relativity, the only things I can know about a black hole from the outside is its mass, its angular momentum, and its electric charge. All other information is unknowable. This is known colorfully as the no-hair theorem. (The idea is that information would represent hairs on this bald ball of a black hole.) I could chuck my die, my cello, and my wristwatch inside, and once they have passed the event horizon nothing about the black hole would give me the slightest hint of what I had thrown in there—or so most physicists believed. But this belief was less of a certainty than it was a conjecture.

In 1974, this fundamental axiom of physics was finally challenged by none other than Stephen Hawking, who suggested that black holes are leaking.

NOT-SO-BLACK BLACK HOLES

Once my die is thrown into a black hole, there is no way of knowing how it will land. At least most scientists thought this must be the consequence of such a concentration of mass warping of space-time. But when Stephen Hawking applied the second law of thermodynamics to black holes, it turned out that they weren't as black as everyone had originally thought.

The second law of thermodynamics states that we are moving from a highly ordered universe into a more disorderly one. What is changing is something called entropy, which is a measure of disorder. Entropy essentially measures the number of different possible scenarios, which in turn suggests how likely it is that any one of them will come to pass. And the second law states that the entropy of the universe is increasing.

A classical example of the increase in entropy is to consider a gas trapped inside a container. If the gas is concentrated in one corner (imagine it has been squashed into the corner), it will, in time, distribute itself across the container. Entropy considers how many different scenarios are possible for the gas. When the gas is trapped in the corner there are fewer possible scenarios than when it can occupy the whole container. Entropy increases as the number of possible scenarios increases. So entropy starts low and rises.

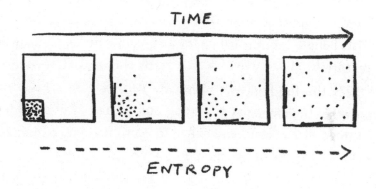

Or consider the more tangible example of an egg falling from a table and smashing on the ground. The highly ordered eggshell becomes a scattered mass of fragments. There are many more ways for the broken shell to be arranged than when it was in one piece wrapped around the egg. Watch a video of this scenario run forward and backward, and it is clear which direction represents the true flow of time. An increase in entropy generally follows the path of time.

This is why entropy is intimately related to the concept of time. It is one of the few things that helps us to know which way the movie is running. Many other laws of physics work perfectly well if I run them forward or backward. Although it is physically possible for an egg to reassemble itself on the table, the decrease in entropy is an indication of how unlikely it is.

You may feel that this move from order to disorder is not always demonstrated on Earth. We have evolved from unicellular creatures in a messy swamp to a state that is in some respects at least more orderly. This seeming violation of the second law of thermodynamics can be explained by the fact that the Earth is taking low entropy from another source—there is a trade going on. The incoming photons from the sun, which are the source of life on Earth, have low entropy. Instead of warming up, the planet radiates heat via electromagnetic waves of lower frequency (and hence energy). So a few high-energy waves from the sun are changed into many low-energy waves emitted by the Earth. The process is a bit like smashing the egg. A single high-frequency photon is absorbed by the Earth, like the egg falling to the ground, and then the Earth kicks out many photons of low energy like bits of eggshell. The Earth benefits by a net decrease in its own entropy, and we witness order emerging out of chaos. But consider the whole system of the Earth and sun, and entropy is indeed increasing just as the second law of thermodynamics demands.

So what happens to a container of gas when it is thrown into a black hole? Or, more intriguingly, what happens to its entropy? Is it lost, contradicting the second law of thermodynamics? Or should we regard the black hole as having its own entropy, which would increase as stuff gets chucked in? That's all very well, but physics tells us that things with entropy have temperature—and things with temperature radiate heat. So it seemed that black holes should be emitting radiation inversely proportional to their mass. And if this is true, then they can't be as black as their name suggests, and should be glowing gently in the night sky.

FUZZY EDGES

This doesn't make sense. How can a black hole be glowing if everything, including light, is meant to be trapped inside? Heisenberg's uncertainty principle tells us that the event horizon is a little fuzzier than the mathematics of general relativity implied. We know that position and momentum cannot both be known precisely, and that time and energy are similarly related in such a way that we can't know both simultaneously. This gives rise to the possibility of quantum fluctuations where matter appears out of nothing.

Hawking realized that the vacuum at the edge of a black hole can experience these quantum fluctuations. This means that a particle and an antiparticle can appear, one with positive energy and the other with negative energy. It may have been this mechanism of creating something from nothing that got the universe up and running. Usually in the vacuum of space, a particle and antiparticle will very quickly annihilate each other. But if this pair should form in such a way that the particle with positive energy is *outside* the event horizon of a black hole and the particle with negative energy is trapped *inside*, something interesting happens. The particle inside gets sucked into the black hole, and since it has negative energy it reduces the mass of the black hole, while the particle with positive energy radiates away from the black hole. The black hole is glowing—it has a temperature, just as we would expect if it had positive entropy.

But hold on. Half the time, the positive-energy particle ends up on the inside rather than the outside. Wouldn't that lead to the black hole increasing in mass? The way to resolve this question is to understand that a negative-energy particle outside the black hole doesn't have the energy to get away, so the net effect of these random fluctuations is an overall decrease in the mass of the black hole over time.

Hawking radiation, as it is known, hasn't been detected emanating from any of the black holes we've identified so far. The trouble is that the math implies that the rate of emission depends inversely on the mass of the black hole. So a black hole with the mass of several suns will be emitting radiation at such a slow rate that it is below the temperature of the cosmic microwave background radiation, meaning that it is impossible to detect against the background noise left over from the Big Bang.

Hawking's proposal provided a mechanism for black holes to fade away, shrinking as time passes. As the mass decreases, the radiation increases, with the result, it's believed, that black holes will eventually disappear with a pop. Hawking predicted the "pop" would be pretty grand, equivalent to the explosion of millions of H-bombs. Others believe it would be more on par with an artillery shell going off.

But this leaves us with something of a puzzle. When I threw the copies of *Private Eye* and *Penthouse* into the black hole, it was believed that this information was lost inside the black hole. But if the black hole disappears, then where does this information go? I could cope with the idea that it was trapped inside the black hole—at least it still existed. But if the black hole eventually disappears, does the information vanish with it? Or is it somehow encoded in the radiation? If I throw my casino die into the black hole, can I somehow tell how it lands from the particles emitted at the edge of the event horizon? Perhaps there is a way to untangle this radiation and retrieve information about everything that has ever disappeared into the black hole. The puzzle of what happens to the information sucked up into the vortex is called the black hole information paradox.

In 1997, Hawking took out another bet, and this time Kip Thorne sided with him. Their bet was with Caltech theoretical physicist John Preskill. They believed that the loss of information was inevitable. Given that it contradicted the theory of quantum physics, Preskill wasn't prepared to concede that information was lost. The wager this time was an encyclopedia of the winner's choice. If it were thrown into a black hole, could the information contained in the encyclopedia somehow be encoded in the new particles being radiated thanks to the uncertainty principle?

In 2004, Hawking dramatically conceded the bet. Based on his latest mathematical calculations he understood that there was a mechanism for the radiation to encode information about the content of the black hole. This was enough for him to change his mind. Preskill collected *Total Baseball: The Ultimate Baseball Encyclopedia*. Hawking later joked, "I gave John an encyclopedia of baseball, but maybe I should just have given him the ashes."

Hawking now believes that the information that plunges into a black hole is actually encoded on the surface of the event horizon and is then imparted back to the particles that are emitted. Weirdly, this surface is two-dimensional, yet it seems to be able to encode information about the three-dimensional space inside. This has led to the idea of the holographic universe: the whole of our three-dimensional universe is actually just the projection of information contained on a two-dimensional surface. Although Hawking conceded the bet, Thorne has stood his ground. He still believes that the information is lost.

Roger Penrose, like Thorne, believes that Hawking conceded too quickly. Once a black hole has radiated away, Penrose believes that information and entropy are lost. For Penrose, this is relevant to the question of why the universe started in such a low-entropy state. The second law of thermodynamics states that entropy is increasing. Entropy is a measure of disorder. The universe is a bit like the egg that started with low entropy, whole and orderly, but then smashed to the ground into a disorderly collection of fragments with high entropy. But this raises some questions. Why was there order to begin with? Where did the order come from? If black holes actually destroy entropy, could they also provide a mechanism for resetting the universe to a low-entropy state?

Penrose had always believed that the Big Bang was a boundary beyond which you couldn't do physics. As we head back toward the singularity of the Big Bang, the equations underpinning our understanding of physics break down. If we accept the description of the Big Bang as an infinitely dense point, on many levels it doesn't make sense to explore the question of what happened before the Big Bang. We can't measure anything that lies beyond it, so we may as well treat it as if it doesn't exist. Or should we? Penrose has changed his mind.

JOINING UP THE PAST AND THE FUTURE

There are stories emerging of what might have happened before the Big Bang, and Penrose tells one of the most remarkable. He proposes that the Big Bang was just one in an infinite cycle of Big Bangs. This isn't the first time such a possibility has been suggested; when it was thought that the universe would end in a big crunch, it seemed a good idea to turn the crunch into the Big Bang of a new era.

But as we discovered in the last Edge, the universe is not contracting but expanding at an accelerating rate, heading for a cold state where no life, galaxies, or even matter will exist, just photons of light. This is what Penrose calls the "very boring era." Even the black holes that will swallow much of the galaxies we see at the moment are believed to leak radiation, so that they too will be spent, disappearing in a final pop and leaving a universe full of photons and gravitons, the hypothetical massless particles thought to mediate the force of gravity.

Penrose admits to being somewhat depressed by this vision of our universe's future: "Good God, is that what we are in for!" he once exclaimed. But then he was struck by the question of who would be around to witness this "overpowering eventual tedium." Certainly not us. The only things around to be bored by events would be the photons and gravitons that alone would make up the universe.

But it turns out that a photon has no concept of time. It lives in a timeless environment. As things approach the speed of light, relativity tells us that time slows down, so much so that by the time something hits the speed of light, clocks stop. But hold on. Light travels at the speed of light. So photons exist outside of time. In fact, in Penrose's scenario, in which all particles with mass will have decayed to massless photons or gravitons, there will be nothing left to mark time, nothing from which to make a clock. Similarly, since time is essential to measuring space, this future universe will have lost all ability to quantify and measure distance—big or small.

Rather than giving in to pessimism, Penrose saw an opportunity. Didn't that sound remarkably like the state of the universe just after the Big Bang? A universe full of energy, but with no matter yet formed. Admittedly, this energy was meant to be concentrated into an infinitely small region to create the conditions for the Big Bang, but if the universe has lost any sense of scale, then couldn't the conditions at the end of our universe be the starting point for a new Big Bang?

These two scenarios—the universe ending in a boring heat death and the universe starting in an exciting Big Bang—can actually be

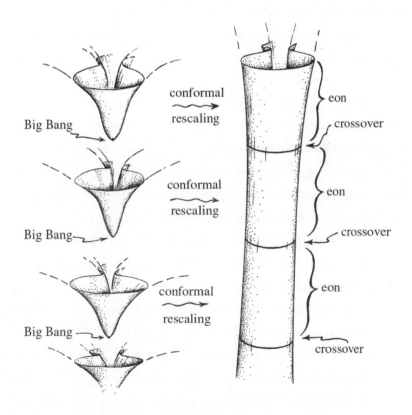

By rescaling the universe at the end of one eon with the beginning of another, we can pass seamlessly from one to another.

seamlessly fused together, like two landscapes whose boundaries match up to create one continuous landscape. Sewing together these scenarios requires that the end of one universe contracts and the beginning of the next expands, so that the two ends fit together, smoothly passing from one to the other. Cold faraway photons become hot close-together photons, initiating a new Big Bang.

Penrose's theory is controversial, and I couldn't find many scientists who thought it was more than a clever idea. And yet when Penrose first revealed that the mathematics of general relativity predicted the existence of singularities in space-time, many dismissed this idea as physically impossible. His current theory of the cycles of time may prove to be wrong, but for me there is something exciting about the possibility of exploring time before the Big Bang.

Penrose calls the period between one Big Bang and the next an eon. Our eon, to his mind, is just one of possibly infinitely many eons that preceded and will follow our own.

One big challenge to this model, and to many cyclical models, is the second law of thermodynamics. This law states that the entropy in the universe is continually increasing. But that means it must be small at the beginning of each new eon. How is entropy reset to such a low state as we go from one eon to the next so that we can run the second law of thermodynamics in each new eon?

The Big Bang occurs in a state of very low entropy. As the universe evolves, entropy increases. How, then, can we move seamlessly into the next eon and reset the entropy? This is why Penrose was unhappy that Hawking conceded the bet about black holes. In Penrose's view, black holes are the mechanism for resetting entropy. All the entropy entering a black hole is lost or subtracted from the whole system, so that by the end of the eon we are at low entropy again, because all the information has been lost in the plethora of black holes that populate the universe. This sets up the conditions for the next Big Bang.

Even if the theory is correct, how can we ever dig back into the period that preceded our own Big Bang to test it? Is "before the Big Bang" a no-go area? Penrose thinks not. Given that the two landscapes must match up, things that happened in the previous eon should have an impact on our own. He believes that the bumping together of black holes toward the closing stages of the last eon will have caused gravitational ripples that passed into our eon. He compares it to a pond into which many pebbles have been thrown. Once the black holes, or pebbles, have disappeared, we will have a pattern of ripples that results from these interacting expanding circles.

Penrose believes that this is something we could look for in the cosmic microwave background radiation left over after our Big Bang. Although the fluctuations across this radiation appear random, perhaps some of them are the result of black holes bouncing off each other toward the end of the last eon.

The trouble is that the cosmic microwave background radiation is notoriously difficult to analyze, partly because there isn't enough of it. You may consider this crazy, given the fact that it makes up the surface of the sphere enclosing the observable universe. And yet, if you are surveying sections of this sphere and you need to take sections that span a ten-degree arc, you are very quickly going to run out of bits you haven't looked at. Although many are skeptical that we will ever see evidence of previous eons in our own, it does raise the exciting prospect that perhaps the question of what happened before the Big Bang is not as unanswerable as we once thought!

A MODERN-DAY GALILEO

I popped into Penrose's office, which is one floor down from my own at the Mathematical Institute in Oxford, to get his take on whether we will ever know what happened before the Big Bang. Penrose, who is now in his eighties, is a great example of how the desire to know never fades. Given that he wrote a book with the modest subtitle *A Complete Guide to the Laws of the Universe*, you may assume that he thinks he knows it all. But he is still asking new questions.

"I used to say that the Big Bang was a singularity so the notion of time doesn't mean anything before the Big Bang. 'Before' is a meaningless question. Don't ask that question. I've heard Stephen Hawking saying not to do so, and I would have agreed. But the view I take now is that you really are allowed to ask that question."

Did this mean that he thought there was no beginning to it all?

"My belief is that it's an infinite sequence of eons."

I asked him whether this might actually be an example of something we could never know. After all, infinity is usually a no-go area for physics.

"We might be able to increase technology and see a few eons back," he said. "But going back to the beginning of them all? Well, there's a candidate for something we may never know. People say that the infinite is unknowable, but in mathematics people use it all the time. We feel completely at home. Well, if not completely, then largely."

When I asked Penrose whether he thought there was any question that by its very nature was unanswerable, he was typically cautious.

"I slightly worry about the concept that there might be anything that will always remain beyond our knowledge. There may be questions that you might not expect an answer to, I suppose. Problems that you think are unsolvable, but then you think of a way around them and get some handle on them. I don't like the term *unknowable*. It just means we're not looking at the thing in the right way.

"You might not think you could ever know what's happening in the middle of the sun, but now people do know what is going on there. Not so long ago, people would have thought of that as an impossible question, I suppose.

"Can you multiply together two incredibly large numbers when the result has more digits than there are particles in the universe and therefore cannot be written down? Does that count as an unsolvable problem? It just seems a tiresome unsolvable problem.

"I think I have a bias—although I wouldn't like to say it's a position I'm taking—a bias in favor of not having absolutely unknowable things." Penrose looked a little concerned. "I hope I haven't disappointed you by saying there are no unknowables."

I suggested that this could be an important mind-set for doing science.

"A problem may be pretty difficult, but somehow you feel there's got to be a solution. I do have that feeling, but I don't know if it is justified. I don't expect to see answers to the big questions in my lifetime, though it would be nice to see a few of the more immediate ones resolved."

I wanted to know which problem he would like to see resolved if he could choose one. Since time is on Penrose's mind, he chose evidence of a time before the Big Bang.

"I'd like to see signals from the previous eon. But we're a long way from that."

So is doing away with the need for a beginning a threat to those who believe in a God that created everything? Penrose laughed as he recalled how he was worried that his proposal might ruffle ecclesiastical feathers, just as his hero Galileo had.

"I did this event at the Vatican and I was a little bit nervous. But then I realized they were honoring Galileo and the invention of the astronomical telescope. So I described my theory of cycles of time and I thought they might be a bit uncomfortable about the Big Bang not being the beginning. Their response: 'No, that's fine . . . God created the whole thing.'"

OUTSIDE OF TIME

The Vatican response points to a question that has always intrigued religious thinkers: What is God's relation to time? This question has become even more intriguing given our modern revelations about time's fluid nature. Einstein's special theory of relativity asks whether you can talk about one event happening before another. From one perspective, event A happens before event B, but Einstein showed that from another perspective event B can occur first.

For religious commentators this poses an interesting challenge. What is God's perspective? For God, did A happen before B, or vice versa? One answer is to respond as the Vatican representative did to Penrose—take God outside of time. Just as God is not located at one point in space, there is no need to locate God at one point in time.

An outsider would look at space-time just as someone on top of a mountain might look down on the lay of the land. But this view of space-time would include the past, present, and future—all of time in one take. Although not phrased in terms of four-dimensional Lorentzian geometry, this is actually the position taken by the fourth-century theologian Saint Augustine of Hippo.

Einstein tried to use this perspective on time to comfort the widow of his friend Michele Besso shortly after his death. "Now he has departed from this strange world a little ahead of me," he wrote to her. "That means nothing. People like us, who believe in physics, know that the distinction between past, present, and future is only a stubbornly persistent illusion." We should regard our existence at this point in time like being in London rather than Paris.

But some theologians have a problem with a God outside of time, because it doesn't leave much room for God to act in the world. If you opt for theism rather than deism, then God needs to have a temporal quality to intervene in the world. Interestingly, although people may argue over the order of events, they won't argue about it if they are causally related. This demands a God that steps in and out of time. But a God that acts in the world is a God that acts in time. So it is very difficult to square a timeless God with a God that acts in the universe.

The question remains—what is this thing called God that is meant to be outside of time? Can anything be outside of time? Actually, there is something that I would regard as timeless: mathematics. And as a timeless thing, it is well suited to the job of sparking the creation of what we observe around us. It is the creator responsible for the equations of quantum physics that gives us space-time and something from nothing. Mathematics has an attractive quality: you don't need to ask who created the mathematics. It is something that exists outside of time and doesn't need a moment when it was created. It just is. Perhaps we should reverse the old adage "God is a mathematician." Perhaps mathematics is the god everyone is chasing. If I replace the word *God* with the word *mathematics* in Thomas Aquinas's attempt at a definition, I think it works quite well: "Mathematics is to be thought of as existing outside the domain of existents, as a cause from which comes everything that exists in different forms."

This is close to theoretical physicist Max Tegmark's idea of the Mathematical Universe Hypothesis, or MUH, which sees our physical universe as an abstract mathematical structure. It's a modern take on Pythagorean philosophy. In the paper in which he proposes this theory, he concludes with the following: "If the MUH is true, then it constitutes great news for science, allowing the possibility that an elegant unification of physics, mathematics and computer science will one day allow us humans to understand our reality even more deeply than many dreamed would be possible."

I probably wouldn't go as far as Tegmark in identifying our physical universe with mathematics. It seems hard, mathematically, to distinguish between two universes in which positive and negative charges are swapped over. The two universes would be physically different, but their mathematical descriptions would be exactly the same. This is an example of something called quidditism, the idea that there is more to the universe than just the relationship between objects—*what* they are (the *quid* is Latin for "what") provides another level of distinction.

If mathematics is eternal and outside of time, you don't need a creator to begin things. Its equations could play the role of something supernatural and godlike. It is not, however, a God that acts in the world. The interesting question is then how many different ways are there to set up a universe out of a set of mathematical equations?

Some argue that simply having an equation that goes from zero unicorns to three unicorns per second doesn't mean unicorns exist. So having equations that allow for the physics of quarks and their interactions with various fields doesn't make the quarks any more real than the unicorns. It's what Hawking refers to as the necessity of understanding how "to breathe fire into the equations." What led to our universe having negative and positive charges set as they are rather than the other way around? Where did the "quid" in quidditism come from?

If there were no universe, no matter, no space, nothing, I think there would still be mathematics. Mathematics does not require the physical world to exist, which to my mind makes it a very strong candidate for the initial cause. Eugene Wigner spoke of the "unreasonable effectiveness of mathematics" to describe its uncanny knack to explain physical phenomena. If physical phenomena were a result of mathematics, it would not be surprising to keep finding mathematical explanations at the heart of our universe.

THOUGHTS ON TIME

There are those who would have us do away with the need to talk of time at all. My watch has been ticking away. It says it is a little past ten in the evening. But what does that mean? Put another watch with the same time in a spaceship, and when it returns it won't match my watch.

Einstein's discoveries reveal that the best I can do is to compare the running of clocks. No one clock measures absolute time. Such a thing does not make sense. If you think about it, this was always the case. How did Galileo discover that swinging pendulums are a good way of measuring time? He was sitting in Mass watching the chandelier in the church swinging in the wind. When he compared it with his pulse, he realized that the time it took to swing from one side to the other did not depend on the angle of swing. But here Galileo was comparing one measure of time with another that he assumed to be constant. The truth is that all these time-keeping devices are relative.

If we go back to physics, although time appears to play a major role, it is possible to rewrite all of the equations without reference to time at all. It seems that because we have such a strong sense of the passage of time, it was the most obvious window through which to observe the world. Books on quantum are all about the evolution of the universe with respect to time. The equation for the trajectory of a ball takes time as input, and outputs where you will find the ball. But none of these books define time, and no physicist has satisfactorily pinned down what we mean by time, so perhaps the best strategy is to eliminate it altogether.

This was the aim of physicist Julian Barbour. Working without an academic position, supporting his family by translating Russian scientific journals, Barbour has developed a theory of physics that removes the need for time. His ideas are articulated in his groundbreaking book *The End of Time*, published in 1999: "Nothing happens; there is being but no becoming. The flow of time and motion are illusions." A number of academic physicists have since taken his ideas very seriously.

But why do I feel that there is something called time flowing along, with me at its mercy? I have the feeling that I can never go back in time, that the future is waiting out there still to happen. I remember the past but not the future. Italian physicist Carlo Rovelli and French mathematician Alain Connes believe that this sensation is a result of incompleteness of knowledge. Called the thermal time hypothesis, it proposes that time is an emergent phenomenon, not a fundamental concept.

If I take any physical system, like the molecules of gas in my room, then generally I don't have complete knowledge of the microscopic state of these particles, but only some overall macroscopic description that could allow for many possible microscopic states. I am reduced to considering the situation statistically because of my incomplete knowledge. Rovelli and Connes are able to demonstrate mathematically how this incomplete knowledge can give rise to a flow that has all the properties we associate with our sensation of time. They believe that time emerges only from this macroscopic consideration of an unknown microscopic system. If you dig deep enough, time will disappear, rather like the way the idea of the surface of a liquid doesn't make sense when you are at the atomic level. Or consider the fact that you can't talk about an atom having a temperature or a molecule of water being wet. Time is not fundamental but a similar emergent property.

That's not to say these things aren't real. Hotness, wetness, time ticking by. I look at my watch. Staring me in the face is 11:55. Another day drawing to a close. Midnight is about to strike, heralding the end of the party, and the feeling that I am one day closer to the moment when I won't be sentient anymore. But why do I feel anything—the passage of time, the pain of a stubbed toe, the pleasurable taste of a fine wine, the excitement of listening to Prokofiev? These things are at the heart of one of the biggest unsolved and possibly unanswerable questions on the scientific books, as we will discover in the next Edge.

SIXTH EDGE

CONSCIOUSNESS

11

My brain? It's my second favorite organ.
> —Woody Allen, *Sleeper*

I'VE DOWNLOADED a smartphone app called Cleverbot, which I'm trying out. It is attempting to convince me that it is human. So I've decided to ask it a few questions to test it out. I've texted a friend of my son's, too, to get a comparison, so here are the responses. Can you tell which is the human and which the chatbot?

Question 1: Do you have a girlfriend?
   Response A: Do you want me to have a girlfriend?
   Response B: Mind your own business.

Question 2: What is your dream?
   Response A: My dream is to become a famous poet.
   Response B: To make lots of money.

Question 3: Are you conscious?
   Response A: If I wasn't I don't think I . . .
   Response B: It's the only thing I'm sure of.

It turns out that the more I play with Cleverbot the more I am training the app to respond like a human. Every conversation I have with the app is banked and tapped into for future conversations, so that my responses will become part of the next conversation that Cleverbot might have.

Although the responses to my questions so far are inconclusive, it's not long before a conversation with Cleverbot reveals it to be rather less than human. But the question of whether my smartphone will ever be clever enough to be conscious of its own existence, or whether I can really tell if my son's friend is truly conscious or is also just a good simulation, is far trickier and goes to the heart of one of the toughest unknowables on the books.

Both responses to my question about consciousness tap into Descartes's famous declaration: "I think, therefore I am." This was his response to the skeptics who doubted whether we could be sure that we truly knew anything about the universe. Originating in Plato's Academy in Athens, the skeptics believed nothing could be known for certain. You think you are currently holding a book or perhaps some electronic device in your hand. But are you sure? I've picked up the die on my table. At least I think I have, but perhaps the whole thing is a dream and there is no book and no die. Perhaps the whole experience is some computer-simulated environment fed into our brains like some scene from *The Matrix*. Descartes in his *Meditations* retorts that in all these scenarios the one thing I can be sure about is my own existence, my own consciousness. However, this "I" could turn out to be one of the ultimate unknowables.

ARE YOU THINKING WHAT I'M THINKING?

There is something scientists call the "hard problem of consciousness." It concerns our internal world—what makes me me? What is it, exactly, that creates our feeling of being? What are the ingredients and mechanism that create consciousness, and how does it emerge? How can I know that my conscious experience has the same quality as yours? Can I ever get inside your head and experience what you are feeling? I feel pain in my head the morning after I've drunk too much, but is that feeling anything like the hangover you experience? I see red when I look at my casino die. You give the color the same name, and we both experience the same wavelength of light. But are you really seeing it how I see it? Does it even make sense to ask that question? When I play my cello, the vibrations of the strings are the same as they reach our ears, but can I ever know that our ears are hearing the same things or that we feel the same thrill of a Bach suite?

We are in a golden age for the question of consciousness. In the Fourth Edge I described how the invention of the telescope allowed Galileo and his contemporaries to probe the edges of the universe. In the Second Edge we saw how the invention of the microscope gave us a tool with which to dig deep into the structure of matter. At the beginning of the twenty-first century we are blessed with new telescopes, telescopes of the mind: fMRI and the EEG scanners allow scientists to peer into our brains and measure the activity associated with our different experiences of red, pain, and cellos.

But even if I can observe exactly the same activity going on inside your head, it doesn't tell me that my conscious experience is the same as yours. Why not? We are built in a very similar way, so why wouldn't it be safe to assume that your internal world is similar? We can use the same principle of homogeneity that is key to doing cosmology. What's happening here is very likely to be the same as what's happening over there. And yet I have only my own conscious experience to go on—one data point. What if my consciousness is uniquely different, yet in all tests my brain appears to behave in exactly the same way as your brain? How could I ever know I was experiencing things differently? Perhaps you are not conscious at all. How can I ever be certain? Language has evolved in such a way that I say "red" to describe all the things that my society declares to be red, yet perhaps I am experiencing something wildly different from your conscious experience of red.

We already know that there are people who have a different quality of conscious experience when they hear a cello or see the number 2. Synesthetes find that other senses are triggered by these experiences. My wife experiences the sensation of very dark red when she sees the number 9 or the letter S. For my favorite composer, Olivier Messiaen, colors were triggered by certain chords. The physicist Richard Feynman had the same Technicolor experience with mathematical equations. These are things that our new telescopes of the mind could explain or at least detect. They could possibly pick up genuinely different brain behavior. But even if all the evidence shows that the person in front of you has a brain that is displaying the same activity as your brain, will you ever be able to know that he or she is truly conscious or just a zombie doing a very good impression of a conscious person? This is the hard question of consciousness, and some believe that it is one whose answer we cannot know.

WHERE IS CONSCIOUSNESS?

Many of the other questions we've been exploring have a rich history and reveal how much we have managed to learn, thanks to our investigations over the centuries. But the problem of consciousness, of getting inside a living brain, is so difficult that, until the dawn of the new technology of the last few decades, it has remained the preserve of theologians and philosophers rather than scientists. That's not to say that scientists haven't been trying to solve the problem.

The question of where my "I" is located has vexed scientists for centuries. I have a strong feeling that my "I" is situated somewhere just behind my eyes. It feels like there is a tiny version of myself sitting there observing the world through my eyes as if it were a cinema screen, making executive decisions about how the body that encases my consciousness will act. There is certainly a feeling that my consciousness is not dependent on my whole body for its existence. If I chopped off my hand, I wouldn't feel as though my consciousness had been split in two. My hand is not me. But how many bits can I chop off before I discover what it is that makes me me?

Not everyone has been convinced that the brain is the seat of our consciousness. Aristotle, for example, thought the brain was simply a cooling agent for the heart, which he believed was the true location of the senses. But others realized that the brain probably was key to who we are. The first-century anatomist Rufus of Ephesus gave us one of the first physical descriptions of the brain. Once the brain is removed from the skull, you first notice that there are three distinct pieces: two halves, which seem to be mirror images of each other, called the cerebrum, and then, sitting underneath, what looks like a tiny version of the brain called the cerebellum.

Cut into the brain and you find holes filled with liquid: the ventricles. In the Middle Ages, scientists thought these cells controlled different mental activities: imagination in the front ventricle, memory in the rear, and reason located in between. Leonardo da Vinci believed that common sense, the thing that fused the five senses into a common experience close to our idea of consciousness, was located in the front ventricle.

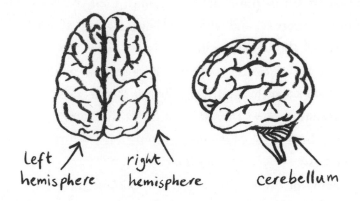

On the left is a top-down view of the brain revealing the two hemispheres. On the right is a view from the left-hand side of the brain showing the cerebellum at the base of the brain.

In contrast, Descartes thought that if we each have a single consciousness, we should be looking for something in the brain that does not have a mirror image but is a single entity. He proposed the pineal gland, a pinecone-shaped gland located at the center of the brain between the two hemispheres that we now know modulates sleep patterns, as the seat of the soul.

Even with a microscope, it is hard to see what it is about this lump of gray matter that could give rise to the complex individual currently reading this book. I'd never seen a brain in the flesh before, and I wondered if meeting one face to face, or brain to brain, might give me new insights. So I visited somewhere with one of the highest concentration of brains in the country. Not Oxford University, but the Parkinson's UK Brain Bank, where I was allowed to hold my first brain, the brain of a man who had recently died and had donated his organ to the tissue bank.

The container in which the brain now sat was labeled C33. But the contents of the vat once had a name like yours or mine. The brain was once home to the hopes and fears, memories and dreams, loves and secrets of the eighty-nine-year-old man who chose to donate his brain to medical science after his death. But where is he now? What was going on inside this brain before the man died that gave rise to his conscious experience, and what is it that has stopped now?

Just from holding the brain, it was no easier to see how it could give rise to our conscious experience. It looked like a huge lump of foie gras. The belief that the ventricles, the liquid-filled holes, were the key to our conscious world turned out to be a red herring; the idea that different regions of the brain were responsible for different functions would, however, turn out to be correct.

By analyzing brains that weren't functioning properly because of damage or lesions, scientists in the nineteenth century began to understand that different parts of the brain are responsible for different functions. For example, the front part of the brain is responsible for problem-solving and decision-making, and social and sexual behavior. The middle portion deals with sensing and perceiving and integrating the data collected by the senses. The back of the brain is like the cinema screen responsible for visual perception, which partly explains the feeling that my consciousness sits at the back of my head watching the film that is my life.

What about the two sides of the brain? There has been much speculation as to the role of each side, but recent research has shown that the brain is perhaps more plastic and flexible than some have thought. That said, we know that the language center is housed principally in the left side of the brain. The nineteenth-century French physician Paul Broca analyzed the brains of patients who had lost the power of speech and discovered that they all had lesions in the same part of the brain, now called Broca's area. Some years later, in 1874, the German physician Carl Wernicke hypothesized that damage to a different area of the brain was causing patients problems with processing language (rather than articulating words). Located toward the back of the left hemisphere of the brain, it is now known as Wernicke's area.

The left side of the brain is also responsible for the number crunching that I do when I'm calculating the probabilities associated with the throw of my casino die. This is where numbers are processed. The right side of the brain is responsible for processing the sound of my cello and listening to music, as well as imagining geometric figures like the icosahedral shape of my cutout universe. To work most effectively, the two halves of the brain are in constant communication through the corpus callosum that divides the two hemispheres. The two hemispheres communicate with each other via nerve fibers that run through this junction. It is a curious design feature, causing quite a bottleneck in communication between the two sides of the brain.

While cutting off a hand certainly isn't going to alter my sense of consciousness, what happens if you cut a brain in two? Since consciousness is a result of brain activity, if I cut the corpus callosum so that the two sides of the brain can't communicate, what happens to my consciousness? Does it split in two?

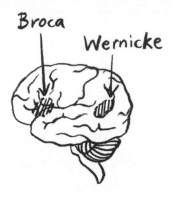

CUTTING CONSCIOUSNESS IN TWO

A surgical procedure called corpus callosotomy, which cuts the corpus callosum, was first carried out in the 1940s on living patients in order to limit epileptic seizures. An epileptic seizure essentially consists of huge simultaneous firing of neurons across the brain. A surge of electrical activity sweeps across the brain, causing the patient to go into seizure. The theory was that cutting the corpus callosum would spare at least one side of the brain from these huge power surges. But what effect did such surgery have on consciousness?

There is good evidence to suggest that the body now housed two distinct consciousnesses. Since each hemisphere of the brain is dedicated to physical behavior on the opposite side of the body, this split consciousness can be observed in the contrasting behavior of the two sides of the body. Extraordinary footage exists of one patient who underwent such a corpus callosotomy, which shows the left side of the body physically attacking the right side. The left side of the body, which is controlled by the right hemisphere of the brain, does not have access to the language side of the brain, which is located in the left hemisphere, so was unable to articulate itself verbally. This frustration seems to have manifested itself in the attack. Eventually, the patient took drugs in order to suppress the right hemisphere and allow the articulate left hemisphere to dominate.

Whether this genuinely reveals a consciousness housed in the right-hand side at loggerheads with the left is difficult to determine. These could be physical responses disconnected from any conscious experience, just as tapping your knee with a hammer produces an involuntary jerk.

Another remarkable experiment that supports the idea that there are two identities at work inside a single body involved counting. A screen was erected between a patient who had undergone a corpus callosotomy and a table on which a number of objects would be placed. The patient could put her right or left hand through a hole in the screen and then use her hand to count how many objects were on the screened-off table.

When the right hand was used, the patient would say out loud the correct number of objects the hand had encountered. But something very curious occurred when the left hand was used. The experimenter would ask the patient to say out loud how many objects the left hand had encountered, and the answer would be completely random and wrong. The language side of the brain (the left), which was saying the number, had no access to the left hand (controlled by the right side of the brain) and was just guessing.

However, when the experimenter asked the patient to put up the number of fingers corresponding to the number of items encountered, she had no problem indicating the correct number. One side of the brain was articulate but had to guess; the other half could only use sign language but had access to the correct answer. This seems less explicable in terms of simple automatic physical response to external stimuli.

Again, the right-hand side might still be able to function intelligently, but perhaps we have here a case of a zombie, without any internal conscious world, that is still able to function like a conscious being. How can we know either way? And in any case, why should we restrict consciousness to the language-generating left hemisphere?

Split brains seem unable to unify information the way that connected brains can. Show the left eye the word *key* and the right eye the word *ring* and the person will say "ring" out loud and can point with his left hand to a picture of a key, but isn't able to unify the two in the concept of a "key ring." But is this an example of two consciousnesses, or simply evidence for the failure of a cut brain to formulate a cognitive connection?

Many patients who have undergone corpus callosotomy are able to function quite successfully. They can drive, work, and function normally in a social setting. So are the two sides actually able to integrate, or is this an example of the two consciousnesses acting successfully in tandem, essentially just two identical copies with one replicating the behavior of the other?

Consciousness takes the multiple inputs that our brain receives from our senses and integrates them into a single experience. In the case of the patients who have undergone corpus callosotomy, the brain is unable to do this. But perhaps there are some brains that even without corpus callosotomy, have difficulty integrating brain activity into a unified experience. Perhaps disorders like schizophrenia or multiple personality disorder are caused by a brain that can't integrate everything into a single voice. The result is that it sounds like there is more than one conscious being inside the brain.

Although these damaged brains helped neuroscientists to understand how certain regions in the brain are associated with certain functions, the real breakthrough in understanding brain architecture came at the end of the nineteenth century, with the work of Spanish scientist Santiago Ramón y Cajal.

SWITCHING ME ON AND OFF

As a young boy Ramón y Cajal wanted to be an artist, but his father had not considered this a suitable profession. He thought medicine was a much worthier cause and came up with an intriguing strategy to spark his son's medical interests: he took him to graveyards in search of human remains. Ramón y Cajal would indulge his artistic disposition by drawing diagrams of the bones they had unearthed, but his father's plan succeeded, and he became more and more fascinated by the human anatomy he was drawing. He received his medical doctorate in 1877.

Ten years later, Ramón y Cajal discovered a way to combine his love of art and anatomy. While a professor at the University of Barcelona, he learned about a new technique using silver nitrate that could reveal the structure of nerve cells. By applying the technique to cells in the brain, he provided some of our first views of the truly extraordinary and complex nature of this organ. Using silver nitrate, he could randomly stain individual cells in the brain to reveal their structure. The results were startling and beautiful.

As Ramón y Cajal stained more and more cells across the brain, he discovered neurons of different shapes and sizes. Using his artistic skill, he filled his sketchbooks with the complex array of structures, documenting them like a butterfly collector. We now know that there are approximately eighty-six billion neurons in the human brain—it would take more than 2,700 years to count them at one per second. Although these neurons come in many different shapes and sizes, as Ramón y Cajal's sketches revealed, they have a very similar underlying structure, consisting of a central cell called the soma, with branches stretching away from the soma called axons and dendrites.

How does a neuron actually work? A neuron gets turned on or fires a little bit like a switch. For example, if my ears pick up a change

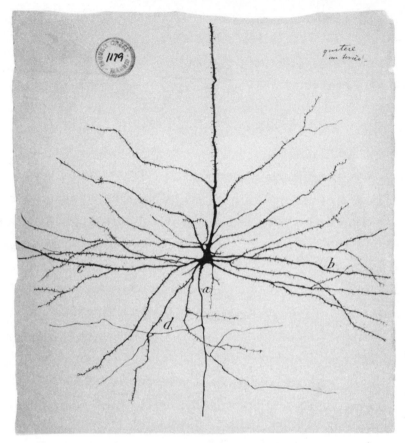

This human retinal cell is one of the first pictures of a neuron. It was the first indication that the brain was not a continuous structure but was made up of discrete, interconnected cells.

of air pressure when I play my cello, this initiates a molecular change in the neuron, causing an electrical current to pass through the cell. This neuron can then talk to another neuron via connections called synapses. Each neuron has one axon coming out of the cell, which is like an electrical wire along which information travels to other neurons. This axon is then connected to the dendrites of other neurons via the synapses. The electrical activity in one neuron can cause a chemical reaction across one of the synapses connected to another neuron, causing it to fire. The brain works very much like a computer or my smartphone with its chatbot app: the neurons are either firing or not.

A few things distinguish the brain from a computer. For one thing, the chemical flow across the synapse must pass a critical rate to trigger the firing of the associated neuron. And the rate of firing of the neurons is as important as the firing itself for the transmission of information. That said, these discrete neuronal cells connected by wires that cause cells to activate or not are extremely suggestive when considering the possibility of creating artificial consciousness in something like my smartphone. But the brain is an extremely complicated network that evolution has wired for us. Each axon can be connected to a thousand different dendrites. This may seem a lot, but given that the brain has eighty-six billion neurons, each individual neuron is actually communicating with only a small fraction of the brain.

This electrical and chemical activity is what gives rise to brain activity. Its loss can give rise to problems and pathologies. The eighty-nine-year-old man whose brain I held at the Parkinson's UK Brain Bank suffered from Alzheimer's disease, which is a direct result of the loss of these neurons and synapses.

We can cut open brains, stain their cells with silver nitrate, and build a static map of the network, but how do we get a picture of the dynamic activity of a living brain? The thing that has revolutionized the study of the brain is the ability to look inside it while it performs certain tasks. The study of the brain has progressed remarkably in recent years thanks to advances in technology.

NEURAL TELESCOPES

The easiest and quickest way to get some idea of brain activity is to use an EEG. When I first had my brain scanned by EEG as part of my ongoing research to understand consciousness, I was a bit nervous. It looked rather like an alien brain-extraction device. In quite a lengthy procedure, which seemed to involve sandpapering my scalp to get a good connection, sixty-four electrodes were attached to the surface of my head, which, while not used to extract my brain, did probe aspects of my thought processes.

Developed by German physiologist Hans Berger in the 1920s, electroencephalography (much easier to call it EEG) records electrical activity along the scalp. It measures the voltage fluctuations resulting from the electrical current flowing within and between the neurons of the brain. Scientists have picked up different kinds of waves corresponding to different brain activity. Synchronized activity of a large number of neurons gives rise to macroscopic oscillations of different frequencies, which correspond to different brain states.

The first-discovered and best-known frequency band is the one known as alpha waves, a synchronized activity of many neurons that gives rise to macroscopic oscillations of eight to twelve hertz. This is a much lower frequency than the notes produced by my cello, the lowest of which vibrates at sixty-five hertz. Nevertheless, it is rather like a musical note singing through the brain. Alpha waves can be detected at the back of the brain during relaxed wakefulness, and increase when our eyes are closed. They're not the only notes the brain plays, and you fire other frequency waves during different periods of brain activity.

- **The slowest are delta waves, at one to four hertz, associated with deep, unconscious, dreamless sleep.**
- **Theta waves are next at four to eight hertz, associated with light sleep or meditation.**
- **Faster than alpha waves are the beta waves at thirteen to thirty hertz, which the brain hits when it is wide awake.**
- **Then come the fast gamma waves at thirty to seventy hertz, which are believed to be most important for the brain's ability**

to create consciousness. Gamma waves are associated with the formation of ideas, language and memory processing, and with various types of learning.

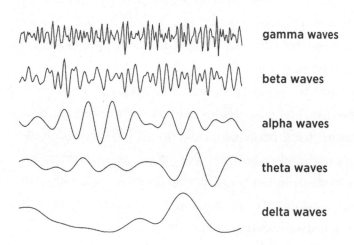

gamma waves

beta waves

alpha waves

theta waves

delta waves

As we make our way through day and night, the brain seems to behave like an orchestra performing a symphony, moving between fast movements and slow movements, with the occasional scherzo when we generate new ideas or encounter new situations.

EEG signals change dramatically during sleep and show a transition from faster frequencies to increasingly slower frequencies such as alpha waves. In fact, different sleep stages are characterized by differences in wave frequency across the brain, which is why these neural oscillations are thought to be linked to cognitive states such as awareness and consciousness. It is when the EEG detects no waveform at all that doctors call a patient brain dead. It seems that the waves inside the brain are the brain's way of synchronizing activity to work in the most efficient manner.

Although the EEG gives scientists fast access to brain activity, the functional magnetic resonance imaging (fMRI) scanner, developed in the 1990s, is the better known of the new tools for seeing what is happening inside the brain. Compared to the compact EEG, the fMRI scanner is a beast. It looks like a space-age sleeping pod, but once it gets going, there is no way you'd be able to sleep. The sound of the magnets at work is so loud that I've needed earplugs when I've been inside a scanner. Although it is not a painful experience, having your brain scanned by fMRI requires keeping still for an unnaturally long time inside the device in order to get a clear picture of what's going on inside your head.

The scanner works by detecting changes in blood oxygenation and flow in response to neural activity. When a brain area is more active, it uses more oxygen. The increased blood flow to the active area can be picked up because oxygenated blood is more magnetic. The fMRI scanner detects this magnetic fluctuation, which can then be used to produce activation maps that show which parts of the brain are involved in a particular mental process.

Since the EEG measures actual brain activity by monitoring the change in electrical behavior, while the fMRI detects a secondary feature (oxygenation), the EEG provides a much better measure of what's actually going on in the brain. But its ability to record brain activity changing over time cannot at present be matched by fMRI scanners. Where fMRI scanners come into their own is in giving a much higher-definition snapshot of the brain. So the combination of the two will give you a pretty good picture of the brain in action. Both EEG and the fMRI scanner can identify the very specific region of the brain that is active when I think about mathematics. But can we use these neural telescopes to see consciousness? The question is whether there is really anything to see.

IS MY CAT CONSCIOUS?

Even if we understand which bit of the brain is firing when we engage in different activities, and how the physics and chemistry of the brain work, it still doesn't give us much understanding of why we have a sense of "I." How can we even approach this kind of question? A common strategy for a mathematician trying to understand what something is, is to try to understand why something else isn't.

Despite my chatbot app's best attempts to convince me otherwise, I do not believe my smartphone is conscious. Neither is the chair I am sitting on. But what about animals? Before he left home, my black-and-white cat Freddie would sit in my office living a life of leisure as I sat scribbling away at my desk. But did he have a sense of self? What about babies? As my children grew up, their brains evolved, and with it their sense of self-awareness changed. So can we identify different levels of consciousness? Are there thresholds in brain development at which different states of consciousness emerge?

Of course, asking my cat about its internal world is a problem. In the late 1960s, animal behaviorist Gordon Gallup was shaving in the mirror, pondering the question of how to test self-awareness in animals, when he was suddenly struck by an idea. He was aware that the face he was looking at in the mirror was his. He wondered which animals would recognize that what they see in the mirror is not another animal but a reflection of themselves.

Perusing the near infinite number of cat videos on the Internet reveals that cats tend to think the image in the mirror is a rival cat in the room. But how can we really tell? Gallup came up with a test to reveal which species recognize themselves in the mirror, which in turn indicates that they have a sense of self.

His test is simple. Introduce the animal to the mirror so that it familiarizes itself with its reflection. (There is fascinating footage of chimpanzees excitedly dancing along with their images in front of the mirror. But do they think they are dancing with another chimp, or are they admiring their own moves?) At some point, the experimenter takes the animal aside and, while wiping its face, surreptitiously places a red mark just below its eye in such a way that it is unaware of the mark and can't see it without looking in the mirror. Gallup wanted to know how the animal would now react to seeing its image in the mirror.

If you looked in the mirror and noticed something strange on your cheek, your immediate reaction would be to touch the mark to investigate it. Gallup's mirror self-recognition test, as it is known, reveals the startling fact that humans are part of a very small group of animals that systematically pass this test of consciousness or self-awareness. The only other species that react in a similar fashion are orangutans and chimpanzees. A third species was added to the list in 2001, thanks to research by Diana Reiss and Lori Marino on bottle-nose dolphins.

Although dolphins have no hands with which to touch their mark, they spent much more time in front of the mirrors in their tanks when they had been marked and were uninterested in marks on other dolphins in their tank, indicating some awareness that the dolphin in the mirror was not just another dolphin. Other individual animals have also passed the test. A clever magpie. An elephant. But not a whole species on a consistent basis.

It is striking that chimpanzees start to fail the test once they reach thirty years old; despite having some ten or fifteen years left to live, it seems that self-awareness comes at a cost. Consciousness allows the brain to take part in mental time travel. You can think of yourself in the past and even project yourself into the future. But at the same time, being aware of your own existence means having to confront the inevitability of your demise. Death-awareness is the price we pay for self-awareness. That is why Gallup believes that later in life chimpanzees prefer to lose their ability to be conscious of themselves. Could dementia in humans play a similar role, protecting aging humans from the painful recognition of their impending death?

Of course, the mirror self-recognition test is a very crude measure of consciousness. It has a bias toward species with highly developed sight. Dogs, for example, do not have good vision but rely on scent to identify other dogs, so you wouldn't expect a dog to pass such a test of self-awareness even if it had an equally well-developed sense of self. Even for those species for whom sight is the primary sense with which to negotiate the world, it is a very rough test of self-awareness. None-theless, it has striking consequences when applied to humans, because we can use it to discover when the brain goes through a transition that allows us to start to recognize the image in the mirror.

I don't think my children, when they were babies, had the same sense of self as they do now. But at what point would they have started to react like the chimpanzees to a mark surreptitiously placed on their faces? It turns out that a child of sixteen months will continue playing in front of a mirror quite oblivious to the new mark, although they might put their hand up to the mirror to investigate the slightly un-usual image.

But place a child of twenty-four months in front of the mirror and you will see her hands reach up for her face at once to explore the strange spot. The two-year-old recognizes the image and thinks, "That's me." Something happens during the brain's development that means we become self-aware, but what it is remains a mystery.

If consciousness emerges in humans between eighteen and twenty-four months, can we ask the same question on a more cosmic scale? Can we figure out when consciousness first emerged in the universe? Surely nothing was sufficiently evolved to qualify as conscious just after the Big Bang. So there must be a moment when the first con-scious experience occurred. Consciousness must then have a different quality to gravity or time.

American psychologist Julian Jaynes, who died in 1997, hypothe-sized that the emergence of consciousness in humans may help to explain the concept of God. With the evolution of consciousness came the creeping awareness of a voice in our heads. Perhaps, Jaynes sug-gested, God was formulated as a way to try to make sense of this emerging internal world.

As you read this now, you can probably sense the words being sounded out in your head. This is part of our conscious world. But those words aren't being sounded out aloud or heard by anyone else. They are part of your conscious world, and yours alone. Jaynes believed that as we evolved and consciousness emerged, the shock of the voice in our heads might well have given rise to the idea of a transcendent intelligence, something that isn't of this world, and this led the brain to interpret it as the voice of God.

This idea of our internal world being close to the transcendent concept of God is central to many Eastern religious practices, including the Vedic tradition. Brahman, the transcendent supreme being of the Hindu religion, is often identified with Atman, or the concept of the self.

Jaynes believed that you can actually date the emergence of consciousness in human evolution. He placed it somewhere in the eighth century BC, between the creation of Homer's *Iliad* and the *Odyssey*. In the *Iliad* there is no evidence in the characters of an internal world, of introspection or consciousness. The characters in the Trojan siege are simply pushed around by the gods. In the *Odyssey*, by contrast, Odysseus is clearly introspective, aware of his own self, conscious in a way the characters in the *Iliad* appear not to be.

TRICKS OF THE MIND

One of the pleasures of reading a book like the *Odyssey* or the *Iliad* is that it immerses you in another world. A good book can make you totally unconscious of your surroundings. The brain is very good at filtering what reaches your consciousness. You don't want to be aware of the whole gamut of sensory input that you receive—it would be overwhelming. But it is striking that your brain can switch between one conscious perception and another without any change in the external input.

My favorite example of this change in consciousness is the brain's reaction to a sketch of my casino die. What do you see?

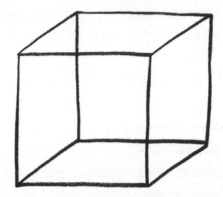

At first it appears to be a cube with one square to the fore. But then, as you stare at it, suddenly the cube flips and it seems as if another square is to the fore. Called the Necker cube, the image hasn't changed, but what you are conscious of has. What has happened in the brain? Is consciousness really just a story that your brain tells about the sensory input it receives through the body's interaction with the world?

Neuroscientist Christof Koch showed me another striking example of how the brain processes visual data in surprising ways. Koch is one of the leading lights in the modern investigation into consciousness. For years, the subject was not something well-respected scientists would say they were studying. It was the preserve of academics in the humanities department, not a subject for scientists in the lab. But when Nobel Prize winner Frances Crick turned his attention from DNA to exploring the question of how the brain generates consciousness, it suddenly became okay for a scientist to take consciousness seriously. Forty years his junior, Koch was Crick's partner in the probing of this problem.

I first met Koch at the top of Mount Baldy just outside Pasadena, home to Caltech, where Koch worked at the time. We'd arranged to meet there, as it was the halfway point on one of the epic runs Koch regularly likes to do. I'm ashamed to say that I declined his offer to join him on a run and cheated by taking the ski lift to the top. Before we grappled with the thorny subject of consciousness, I couldn't help but be distracted by a tattoo Koch sports on his shoulder. It seemed to be the rainbow bitten-apple logo used by the Apple computer company.

"I had it done in 2000 while I was on an archeological dig in Israel with my son," he told me. "The Apple computer is one of the most beautiful and elegant artifacts of the twentieth century. A perfect marriage of form and function."

I think he believes that the Apple computer will be as significant to understanding today's culture as the pots he was digging out of the ground with his son are to appreciating Herod's Caesarea in 20 BC. Koch talks a lot about his Apple computer and is constantly wondering whether one day it might be conscious and talk back to him. His other love is his dogs, which he believes have a much higher level of consciousness than we give them credit for. It was his belief in animal consciousness that led him to become a vegetarian.

"Because it is likely that mammals can consciously experience the pains and pleasures of life, can be happy or sad, we should not be eating their flesh," he explained. Then he reflected, "It was difficult to immediately follow this growing realization with action—the taste of meat is very deeply ingrained."

Koch's area of neurological expertise is vision, and with a piece of A4 paper he had in his pocket he illustrated how vision gives us some interesting insights into how things reach our conscious minds. Koch gave me the A4 paper and asked me to roll it up into a cylinder to make a telescope. Then he told me to bring the telescope up to my right eye and, keeping my left eye open, to hold my left hand open about four inches from my left eye.

"Now look at the mountains over there. What do you see?"

As I looked across at the view I could not stop laughing. I seemed to have a hole in my hand!

Koch explained that this was the case because the brain was try-ing to process two seemingly inconsistent pieces of information. So what reached my conscious brain was a fusion of what it thought I might be interested in. I was seeing part of my hand from the visual information entering my left eye, but also the small circle at the center of the telescope from my right eye. Superimposed, it appeared that I had a hole in my hand. Koch believes that how the brain decides what to make my consciousness aware of could help us to understand more about consciousness itself.

Can we use our new neural telescopes to see what happens when our brain becomes conscious of one thing or another? According to Koch's research, it appears that the retinal nerves are not sending dif-ferent information, so the change in our conscious experience is hap-pening further down the line in the brain. The trouble is that the fMRI and EEG scanners are too crude to pick up changes as subtle as the flip from one view of the Necker cube to the other. But in 2004 Koch, together with his team at Caltech, spotted an opportunity to ask an individual neuron questions about what stimulated it to fire, which led to the discovery of some curious neuronal activity.

THE JENNIFER ANISTON NEURON

Epileptic seizures can be caused by a miswiring of the synapses, or by scar tissue that triggers the synchronized firing of neurons across the brain, a bit like the chain reaction of an atomic bomb. In some cases, the removal of a small section of the brain can prevent such seizures by removing the trigger that sets off the cascade.

The procedure is daunting. To locate the source of the seizure, and so as not to remove too much tissue, around twenty electrodes are inserted into the soft tissue of a patient's brain through holes drilled in the skull. It sounds like some medieval torture regime, but is actually a procedure that is very much part of current medical practice. Each electrode has tiny wires, as thin as a hair, that are wired up to a region of about ten to fifty neurons. If any of these neurons fire, the electrical activity is picked up by the electrode. The surgeon then waits for the patient to have a seizure, at which point a recording can be made of the neurons firing across the brain. A mathematical analysis of the data then pinpoints the potential source of the surge.

But a seizure can take some time to occur, and in the interim these wired-up patients sit on the ward twiddling their thumbs. The Caltech team saw an opportunity. Why not ask them questions and see which ones made their neurons fire? The region of the brain that is the most common source of seizures is also where memories are encoded. Since this is where the bulk of the electrodes were attached, they decided to see what electrical activity could be picked up when memories were fired. To trigger these memories, the patients were shown a range of different pictures.

It took some probing, but the results were striking. In one instance, neurons would fire only when the patient was shown pictures of the actress Jennifer Aniston. It didn't matter if she was dressed differently or her hair was a different color; the neuron seemed to fire in recognition of the concept of Jennifer Aniston. Even when the patient was just shown the name, the neuron would fire.

In one sense, the discovery was not surprising. To encode memories, ideas, and concepts, the brain has to find some way to convert the data into neuronal activity. Just as a digital photo of Jennifer Aniston translates her image into a series of 0s and 1s, our brains take the sensory data that bombards us and decide whether Jennifer Aniston is important enough to be encoded as a concept. If she is, then the synapses that are currently connected to that neuron will strengthen and survive, ensuring that the neuron fires when the brain receives the associated visual information. Each concept will have its own characteristic synaptic chain of neuronal activity. I particularly liked the patient whose neuron fired whenever he was shown an image of Pythagoras's theorem—a much more discerning neuron!

What was surprising was how selective the neuron was in its firing. It didn't seem to be interested in any other images. Koch told me that this neuron was certainly not the only one involved in encoding the idea of Aniston or Pythagoras. That would be extremely inefficient. The experiment was limited in the number of neurons it could question, and just like the digital 1s that are switched on to record a digital photograph in my smartphone, the brain has a selection of neurons that fire across it to encode the concept of Aniston. But Koch believes that a remarkably small number are involved in the encoding, in the hundreds or perhaps thousands, rather than the millions of bits that would be required to encode such images on a computer.

This encoding helps explain something called qualia, a quality or property of things as perceived or experienced by someone, like the color red. It doesn't matter whether you are looking at a tomato, a fire truck, or a drop of blood, you will experience a sensation of redness. The challenge is to determine whether your qualia are anything like mine. Do animals or computers experience qualia?

Koch's results suggest that we can think of what happens when we are shown a tomato as billions of 0s and 1s lighting up in the brain as the concept is recognized. These different code words can be thought of as points that can be lined to make out shapes, like crystals. Is there something common to the shapes of the encodings of a tomato and a fire truck that means that we experience the feeling of red? Could this be the source of synesthesia? Perhaps in some brains the encoding of the qualia for red has a shape that is very close to the shape that encodes the concept of the number 7, so much so that when the brain recognizes the concept of 7 it fires the sensation of red.

Returning to my paper telescope and the hole in my hand, Koch told me you can modify the experiment a little to test whether being conscious of something is enough to make these concept neurons fire. Suppose we have two paper telescopes, one attached to each eye. At the end of one telescope I place a picture of Jennifer Aniston. Sure enough, the Jennifer Aniston neurons in my brain fire. But now I flash an image of Pythagoras's theorem at the end of the other telescope. The brain now has to choose which image to prioritize, or which one will reach your consciousness.

What tends to happen is that the image of Jennifer Aniston disappears in deference to the new image. Pythagoras's theorem takes precedence, so that the participant is no longer conscious of Jennifer Aniston, despite the fact that the brain is still receiving all the same Aniston-related input.

Can we observe a change in brain activity corresponding to the change in consciousness? Experiments similar to those conducted on epileptic patients have been run on monkeys, and these indicate that the neurons that fire in recognition of one concept stop firing when a second image is introduced. So there is something in the electrochemical activity of the brain that contributes to what we are conscious of. At some level this seems obvious, but suppose you think that consciousness is the result of an as-yet undiscovered force. Then, if the neurons keep firing even when we are no longer conscious of the image of Jennifer Aniston, this other force could be controlling why we are suddenly unaware of the image.

This ability to switch our conscious experience on and off lies at the heart of many magicians' acts. I vividly remember an evening with Fellows at the Royal Society when we were shown a video by Richard Wiseman, a professor of psychology at the University of Hertfordshire. The video showed two teams passing two basketballs between them. "I want you to count how many passes of the ball the team in black makes," Wiseman instructed us. In a classic illusionist's move, we were sold a red herring to ensure that we all concentrated our attention on the task at hand: "Men and women tend to count the passes differently."

The video played and we all counted away. When the video finished, Wiseman asked who had counted seventeen passes. Hands went up. Eighteen passes? Some other hands went up. "And who in the room saw the man in the monkey suit walk to the center of the court and bang his chest before walking off?" What! I hadn't seen anything of the sort. I thought he was having us on—until I saw two people with their hands up. They hadn't bothered to concentrate on counting the passes, and, with their attention free to wander, they had registered the man in the monkey suit. It was an important message to the scientists sitting in the room at the Royal Society: focus on one thing too much and you will miss the monkey right in front of your eyes.

This sense of things floating in and out of consciousness is familiar to me as a research mathematician. I have often felt that the work I do at my desk sows the seeds of ideas that only come to fruition in a flash once I've moved on to other things. The mathematical discovery I am most proud of certainly seemed to come to me in this way. I had been working away all day at the Max Planck Institute in Bonn, trying to come to grips with a thorny problem I was trying to solve with a colleague. Nothing seemed to give. That evening, when I'd switched off from consciously trying to solve the problem, as I tried to phone my wife in London, I suddenly "saw" a new symmetrical object with strange and unexpected properties. I quickly scribbled down on my yellow legal notepad the idea I'd suddenly become conscious of, and to this day I find it quite extraordinary how this thing appeared in my mind, an object that had never been conceived of before that moment. It was as if this new mathematical object was thrust into my conscious brain, having been honed in my subconscious, accompanied by a chemical rush to make sure I didn't miss it. But still, what happened so that I had that experience is a mystery.

OUT-OF-BODY EXPERIENCES

Koch's paper telescope reveals how easy it is to confuse our conscious senses and change how we experience our environment. But there are even more spectacular tricks you can play on the brain. By messing with several senses at the same time, it is possible to alter quite dramatically our consciousness such that we can shift our sense of "I" outside our bodies completely.

Search for "the McGurk effect" on YouTube and you will find a video that illustrates the extraordinary power of the brain to create conscious experiences that aren't really there.

At first you see a face that appears to be saying "Fa . . . Fa . . . Fa." But then you are told to close your eyes, and the sound suddenly turns into "Ba . . . Ba . . . Ba." This was actually the sound that was being made all along, but because you see a face making the movements associated with the sound "Fa," your brain is confused by the contradictory input. The brain always seeks one integrated conscious experience, and vision generally trumps sound when it comes to integrating the different senses. Even when you know what is going on, it is very difficult to make your brain hear "Ba" as you watch the mouth make the sound of "Fa." The brain is a pattern searcher, eager to impose structure on the cacophony of information it is being bombarded with. When the information is ambiguous, it must choose.

These illusions offer a potent warning to all of us trying to solve the deeper mysteries of the universe. We have no privileged access to reality. Our interaction with our environment is constructed from the information our brains receive, and we build a plausible representation of the external world. This mixing of the senses has some very strange implications for the location of our consciousness.

An experiment I participated in at the Karolinska Institute in Sweden really made me question where "I" was. Developed by Henrik Ehrsson in 2007, the experiment succeeded in making me feel as though my consciousness was actually located in someone else's body. Ehrsson's inspiration stems from a famous experiment called the false hand illusion, where the participant's hand is placed out of view, behind a screen, and a false rubber hand is placed in view and joined via a sleeve to the participant's body. Initially, the participant sees it for what it is: a false hand. But if the experimenter starts stroking both the false hand (in view) and the real hand (out of view) in such a way that the actions are synchronized, there's a strange shift.

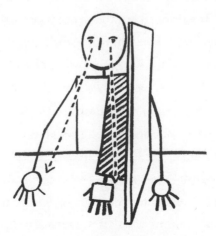

The combined visual and tactile stimulus is ambiguous, so the brain tries to make sense of what is going on. Vision again dominates, and the participant begins to identify with the false hand, so much so that when a hammer is taken to the false hand, the participant very often reacts as if he or she is being attacked. The combination of the sense of sight and the sense of touch has tricked the brain into identifying with the false hand. What's striking is the plasticity of the brain, which has reprogrammed itself within only a few minutes to identify with a false hand.

At the Karolinska Institute this combination of visual and tactile input has been taken one step further. Many recent advances in neuroscience are due to the development of new technology—in this case, virtual reality goggles. I was asked to wear the goggles while Ehrsson donned what looked like an academic mortarboard with two cameras on top. These cameras were to become my eyes. By taking the video feed from the cameras on top of Ehrsson's head, I was given Ehrsson's perspective. Nothing too strange there. It was when he asked me to shake his hand that things became somewhat surreal.

As we shook hands, Ehrsson asked me to squeeze his hand in a regular rhythm that he matched, and this combination of vision and touch caused me to identify more and more with the arm that I could see supposedly emanating from my body. The weird thing was that the arm I was looking at belonged not to me but to Ehrsson. The identification was so strong that when Ehrsson produced a knife and drew it across his hand, I reacted as if my hand were about to be cut. Since my visit to Stockholm, Ehrsson has pushed this illusion even further, making participants feel as if their consciousness is located in a Barbie doll.

A number of recent science fiction films have explored this idea of out-of-body experience, including *Avatar* and *Surrogate*. Having taken part in Ehrsson's experiments, I wonder just how long it will be before I can send an avatar to a concert or to the top of Everest and, by feeding my body at home with all the sensory input experienced by the avatar, feel as though I am inside the concert hall or standing on top of the world.

These experiments get to the heart of what is called the mind–body problem. Is consciousness separate from our physical body? It certainly appears to be created by it, but should we regard it as something distinct? Some have argued that the mind is genuinely independent of the body. This was Descartes's view, one that he shared with many Enlightenment thinkers.

Descartes was a dualist; he believed that the mental world was separate from the physical one. This dualism lies at the heart of many people's faith. Some have described this "other thing" as the soul. But how independent is that other thing from the body to which it is connected? Ehrsson's experiments challenge our sense of the spatial unity of mind and body. But our sense of self still seems very dependent on the body, as it is by manipulating sensory input that we are able to play with the nature of our conscious experience.

The concept of "emergence" has become popular recently in many fields of science. The term was coined to express how some things arise from more fundamental things and yet are themselves fundamental and irreducible. Wetness is the classic example of an emergent phenomenon. One molecule of H_2O is not wet; wetness requires that you put many molecules together before it appears. Emergence, when applied to consciousness, is an attempt to find a middle ground between hardcore reductionism and Descartes's spooky dualism.

Many neuroscientists now speak of consciousness as being similar to the wetness of water. Consciousness is an emergent phenomenon in the sense that it is a higher-level property of a system triggered by neuronal activity happening at a lower level. But that doesn't really explain what this higher-level thing really is.

Some have argued that the concept of emergence is really just a way for chemists and biologists to claim that their subject is distinct from, and not merely a corollary of, physics and mathematics. A hardcore reductionist will tell you that a chemical reaction or a biological process is just a mathematical equation describing physics in action.

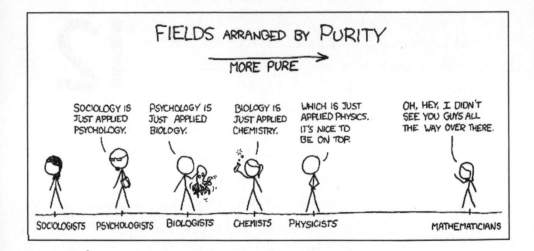

Isn't the app on my phone that is trying to convince me it is conscious really just a mathematical code? When I ask Cleverbot whether it is just a piece of mathematics, it comes back with an interesting answer:

"I might be. Are you?"

12

What we are today comes from our thoughts of yesterday and our present thoughts build our life of tomorrow: our life is the creation of our mind.

—*The Dhammapada*

S OME YEARS AGO, my wife was in a coma following an accident. I spent many hours in the hospital wondering whether "she" was still inside, or whether this was just a body with no consciousness. Fortunately she woke up after two weeks, but what about those who don't? For some, the body has become a collection of cells with no conscious experience—this is what we call the vegetative state. But for others there is the terrifying scenario of a consciousness locked in a body that won't respond. So is there a way to ask the brain whether it is conscious in these situations? Amazingly, there is, and it involves playing tennis—or at least imagining that you are playing tennis.

Think of yourself playing tennis. Big forehands. Overhead smashes. Try to conjecture yourself in the middle of a point. To do so, you have to make a conscious decision to participate. But once you start imagining, an fMRI scan can pick up the neural activity that corresponds to the motor activity associated with the shots you are picturing yourself playing.

In early 2006, British neuroscientist Adrian Owen and his team discovered that we could use an fMRI to see patients making conscious decisions. They were investigating a twenty-three-year-old woman who had been diagnosed as being in a vegetative state after a traffic accident. When the patient was instructed to imagine that she was playing tennis, Owen saw, to his surprise, the region corresponding to the supplementary motor cortex light up inside the scanner. When asked to imagine taking a journey through her house, a different region called the parahippocampal gyrus, which is needed for spatial navigation, was activated. It was an extremely exciting breakthrough, and the doctors quickly rediagnosed this young woman as conscious but locked in. Unfortunately, they were not able to do much to improve her condition and she died five years later.

Many other patients who had previously been thought to be in a vegetative state have now been rediagnosed as locked in. This makes it possible for doctors and, more important, family members to talk to them and solicit responses. A patient can indicate the answer "yes" to a question by imagining that they are playing tennis. Conversations are a bit more tricky, but the ability to trigger and measure a neural response has given us an extraordinary conscious-o-meter.

I recently volunteered to be participant number 26 in a research project in Cambridge seeking to explore the moment when anesthesia knocks out consciousness. There have been some terrible examples of patients who can't move their bodies but are completely conscious during an operation, and scientists want to understand how much anesthetic is needed to really knock out a patient's consciousness.

So I found myself lying inside an fMRI scanner while plugged into a supply of propofol, the drug Michael Jackson had been taking when he died. After a few shots, I must admit I could understand why Jackson had become addicted. It had an extremely calming effect. But I was here to work. With each new dose, I was asked to imagine that I was playing tennis. As the dose was increased, the research team could assess the critical moment when I lost consciousness and stopped playing imaginary tennis. I was very curious to know just how much more propofol was needed to go from knocking out my body so that it was no longer moving to the moment when my consciousness was turned off.

The ability to question the brain even when the rest of the body is unable to move or communicate has allowed researchers to gauge just how much anesthetic you need to temporarily switch off someone's consciousness so that you can operate on them. This tennis-playing conscious-o-meter leads us also to the close relationship between consciousness and free will: you must choose to imagine playing tennis. But how much can we really choose? Recent experiments looking at the brain in action have fundamentally challenged our understanding of free will.

AM I IN CONTROL?

I'm sure you thought you were asserting your free will when you decided to read this book. You probably feel it was a conscious choice to pick up the book and turn to this page. But free will, scientists are learning, may be just an illusion.

Unlike the discovery of the quark or the expanding universe, it is possible to witness much of the cutting-edge research into consciousness as it is conducted, even to be part of the research.

I hope fMRI scanners are not bad for you, because in my quest to understand how the brain shapes my conscious experience here I was in another city (Berlin this time) and another scanner. Lying inside, I was given a little handheld console with two buttons: one to be activated with my right hand, the other with my left. John-Dylan Haynes, the neuroscientist who developed the experiment, asked me to go into a Zen-like state of relaxation, and told me that whenever I felt the urge I could press either the right-hand or left-hand button.

I was then given a pair of goggles with a tiny screen on which a random stream of letters was projected. I was asked to record which letter was on the screen each time I decided to press the right or left button.

The fMRI scanner recorded my brain activity as I randomly pushed the right or left button. What Haynes discovered is that, by analyzing my brain activity, he could predict with a surprising degree of accuracy which button I would press six seconds before I myself became aware of which one I was going to choose. Six seconds is a very long time. My brain decides which button it will instruct my body to press. Then one elephant, two elephants, three elephants, four elephants, five elephants, six elephants—and only now is the decision sent to my conscious brain.

Haynes can see which button I will press because a region in the brain lights up six seconds before I do so, preparing the motor activity. A different region of the brain lights up according to whether the brain is preparing the left or right finger. Haynes can't predict with one hundred percent certainty, but his predictions are accurate more than sixty percent of the time, and he believes that with more refined imaging, it may be possible to get close to one hundred percent accuracy.

It should be stressed that this is a very particular decision-making process. If you are in an accident, your brain makes decisions in a split second, and your body reacts without the need for any conscious brain activity to take place. Many processes in the brain occur automatically, so that the brain is not being overloaded with routine tasks. But whether I choose to press the right or left button is not a matter of life or death. I am freely making a decision with the luxury of time.

The brain seems to prepare the decision far in advance, but it is still not clear where the final decision is being made. Maybe I can override the decision my brain has prepared for me. If I don't have "free will," some have suggested that perhaps I at least have "free won't." In this experiment, the choice of button is immaterial, so there is little reason not to go with the decision my unconscious brain has prepared for me. But what if my brain tells me to do something that I know to be bad for me? What if it tells me to reach for heroin?

The experiment suggests that consciousness may be a secondary function of the brain. We already know that so much of what our body does is totally unconscious, but what we believe distinguishes us as humans is that our consciousness is an agent in our ultimate actions. At some point, we decide. But what if it is only way down the chain of events that our consciousness kicks in? Is our consciousness of a decision just a chemical afterthought with no actual influence on what we do? What implications would this have legally and morally? "I'm sorry sir, but it wasn't me who shot the victim. My brain had already decided six seconds before that my finger was going to pull the trigger." Be forewarned: biological determinism is not a strong legal defense!

The button pressing was of no consequence, and this may skew the outcome of the experiment. Haynes followed his experiment with a slightly more intellectually engaging activity. Participants were now asked whether to add or subtract a number to numbers shown on a screen. Again, the decision was suggested by brain activity four seconds before the conscious decision to act.

But what if we take the example proposed by the French philosopher Jean-Paul Sartre of a young man who must choose between joining the resistance or looking after his grandmother. Perhaps in this case conscious deliberation will rescue free will from the fMRI scanner. Maybe the actions performed in Haynes's experiments are too automatic. Free will is replaced by the liberty of indifference.

The preconscious brain activity may well be just a contributing factor to the decision that is made at the moment of conscious deliberation. It does not necessarily represent the sole cause, but helps inform the decision—here's an option you might like to consider.

If we have free will, one of the challenges is to understand where it comes from. I don't think my smartphone has free will. It is just a set of algorithms that it must follow in a deterministic manner. Just to check, I asked my chatbot app whether it thought it was acting freely, making choices. It replied, rather cryptically, "I will choose a path that's clear, I will choose free will."

I know that this response is programmed into the app. It may be that the app has a random-number generator that ensures its responses are varied and unpredictable. But randomness is not freedom—those random numbers are being generated by an algorithm.

Free will is one of the things I cling to because I believe it marks me out as different from an app on my phone. I think this is why Haynes's experiment left me with a deep sense of unease. Perhaps my mind is just the expression of a sophisticated app at the mercy of the brain's biological algorithm.

Alan Turing was one of the first to question whether machines could ever think intelligently. He thought a good test of intelligence was to ask, if you were communicating with a person and a computer, whether you could distinguish which was the computer. This test, now known as the Turing test, was the one I was now putting Cleverbot through. If a computer can pass itself off as human, shouldn't I call it intelligent?

But isn't there a difference between my brain and a machine following instructions? If I type a sentence in English into my smartphone, the apps on board are fantastic at translating it into any language I choose. But no one thinks the smartphone understands what it's doing. The difference perhaps can be illustrated by an interesting thought experiment called the Chinese Room, devised by philosopher John Searle of the University of California. It demonstrates that following instructions doesn't prove that a machine has a mind of its own.

I don't speak Mandarin, but imagine that I was in a room with an instruction manual that gave me an appropriate response to any string of Chinese characters. I could have a very convincing discussion with a Mandarin speaker without ever understanding a word of my responses. In a similar fashion, my smartphone appears to speak many languages, but we wouldn't say it understands what it is translating.

It's a powerful challenge to anyone who thinks that just because a machine responds like a conscious entity, we should regard that as a sufficient measure of consciousness. Sure, it could be doing what a conscious person would do, but is it *really* conscious? Then again, what is my mind doing when I'm writing down words? Isn't it just following a set of instructions? Could there be a threshold beyond which we would have to accept that the computer understands Mandarin, or that the algorithm should be regarded as having consciousness? But before we can program a computer to be conscious, we need to understand what is so special about the algorithms at work in a human brain.

THE SCIENCE OF SLEEP

One of the best ways to tease out the correlation between consciousness and brain activity is to compare our conscious brain with our unconscious brain. Is there a noticeable difference in the brain's activity? We don't have to wait for a patient to be in a coma or under anesthetic to make this comparison, as there is another state we go into every day—or rather every night—when we lose consciousness: sleep. The science of sleep may give us the best insight into what the brain does to create our experience of consciousness.

So I went to the Center for Sleep and Consciousness at the University of Wisconsin–Madison, where experiments conducted by neuroscientist Giulio Tononi and his team have revealed strikingly different brain behavior during waking and dreamless sleeping.

In the past it was impossible to ask the sleeping brain questions. But transcranial magnetic stimulation, or TMS, allows scientists to infiltrate the brain and artificially fire neurons. By applying a rapidly fluctuating magnetic field, the team can activate specific regions of the brain when their subjects are awake and, more excitingly, when they are asleep.

I was a bit nervous about the idea of someone zapping my brain. After all, it is my principal tool for doing my job. Scramble that and I'm in trouble. But Tononi assured me it was quite safe. Using his own brain, one of his colleagues demonstrated how he could activate a region connected with the motor activity of the hand. It was quite amazing to see that zapping this region caused his finger to move. Each zap threw a switch in the brain that made the finger move. Given that Tononi's colleague seemed unaffected by the zapping, I agreed to take part in the experiment.

Tononi and his team started by applying TMS to a small region of my brain when I was awake. Electrodes attached to my head recorded the effect via EEG. The results revealed that different parts of the brain respond at different times in a complex pattern that feeds back to the original site of the stimulation. The brain, Tononi told me, is a complex integrated network. Its neurons act like a series of interrelated gates. One neuron may fire if most of the neurons connected to it are firing. Or it may fire only if one connected neuron is on. What I was seeing on the EEG was the logical flow of activity caused by the first zap.

The next part of the experiment required that I fall asleep. The room was plunged into semidarkness to help me nod off. I'd just stepped off a transatlantic flight, so my body clock was rather unclear as to what time it really was. I was exhausted and ready for sleep, but a head covered in electrodes wasn't going to make it easy. Once I would be in a deep "stage four" sleep, Tononi and his team would apply TMS to my brain in exactly the same location, stimulating the same region. Essentially, the TMS fires the same neurons, turning them on like switches, to investigate how the structure of the network changes as the brain goes from conscious to unconscious. This part of the experiment proved to be extremely difficult for me. I am an incredibly light sleeper. Having a head covered in electrodes and knowing that someone was about to creep up and zap me once I'd dropped off isn't very conducive to deep sleep. Despite depriving myself of coffee for the whole day, I failed to get past a light "stage one" fitful sleep.

Instead, I had to content myself with the data from another more cooperative patient. The results were striking. Electrical activity does not follow the same path when we are asleep, as it does when we're conscious. It's as if the network were down. Any activity is very localized. The exciting implication is that consciousness may be related to the way the network feeds information back and forth across the brain.

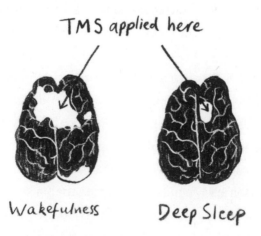

TMS applied here

Wakefulness   Deep Sleep

Having failed to fall asleep in his lab, I was taken by Tononi to his office, where he promised he would make up for my caffeine deficiency by crafting the perfect espresso from his prized Italian coffee machine. But he also had something he wanted to show me. As we sat down to the intoxicating aroma of grinding coffee beans, he passed me a paper with a mathematical formula penned on it.

My interest was immediately piqued. Mathematical formulas are for me like Pavlov's bell. Put one in front of me and I'm immediately hooked and want to decode its message. But I didn't recognize it.

"It's my coefficient of consciousness," he said with a suggestive smile.

Consciousness in a mathematical formula . . . how could I resist?

Tononi explained that as a result of his work on the sort of networked behavior that corresponds to conscious brains, he has developed a new theory of networks that he believes are conscious. He called it integrated information theory, or IIT, and it includes a mathematical formula that measures the amount of integration and irreducibility present in a network, which he believes is key to creating a sense of self. Called Φ (that is, phi, the twenty-first letter of the Greek alphabet) it is a measure that can be applied to machines like my smartphone as much as it can the human brain, and it offers the prospect of a quantitative mathematical approach to what makes me "me." According to Tononi, the larger the value of Φ, the more conscious the network.

A conscious brain is like a network with a high degree of connectivity and feedback. If I fire neurons in one part of the brain, it results in a cascade of cross-checking and information fed across the network. If the network consists of only isolated islands, then it appears to be unconscious. Tononi's coefficient of consciousness is a measure of the extent to which the network is more than a sum of its parts.

Tononi believes it's the nature of this connectivity between the neurons in the brain that is important. It isn't enough that there should be high connectivity across the network. If the neurons start firing in a synchronous way, this does not seem to yield a conscious experience. (We can find synchronous brain activity during deep sleep, and seizures are often associated with the highly synchronous firing of neurons across the brain.) It seems to be important that there is a large range of differentiated states. Too many patterns or symmetries in the wiring can be a problem, as they can sometimes lead to an inability to differentiate experiences. Connect a network up too much and the network will behave in the same way whatever you do to it.

Tononi's coefficient tries to capture one of the extraordinary traits of our conscious experience: the brain's ability to integrate the vast range of inputs our body receives and synthesize them into a single experience. Consciousness can't be broken apart into independent experiences felt separately. When I look at my red casino die I don't have separate experiences: one of a colorless cube and another of a patch of formless color. Tononi's coefficient also tries to quantify how much more information a network will generate when it is complete than when it is cut into disconnected pieces, which is what happens to the brain during deep sleep.

Tononi and his colleagues have run interesting computer simulations on "brains" consisting of eight neurons, connecting them in different ways to see which network has the greatest Φ. The results reveal that you want each neuron to have a connection pattern with the rest of the network that is different from that of the other neurons. But at the same time, you want as much information as possible to be able to be shared across the network. If you consider any division of the network into two separate halves, you want the two halves to be able to communicate with each other. It's an interesting balancing act between overconnecting and cutting down differentiation versus creating difference at the expense of communication.

The nature of the connectivity is also important. Tononi's team has created two networks that are functionally equivalent and have the same outputs, but in one case the Φ count is high, because the network feeds back and forth, while in the other the Φ count is low because the network is set up so that it only feeds information forward (the neurons can't feed back). Tononi believes this is an example of a "zombie network": a network that outputs the same information as the first network but has no sense of self. Just by looking at the output, you can't differentiate between the two networks, but according to Tononi's Φ, the zombie network has no internal world.

It's encouraging to know that the brain's thalamocortical system, which is known to play an important part in generating conscious-ness, has a structure similar to networks with a high Φ. Contrast this with the network of neurons in the cerebellum, which doesn't create consciousness. Called the little brain, the cerebellum is located at the back of the skull and controls things like balance and our fine motor control. It accounts for eighty percent of the neurons in the brain, and yet, if it is removed, although our movement is severely impaired, we do not have an altered sense of consciousness.

In 2014, a twenty-four-year-old woman in China was admitted to the hospital with dizziness and nausea. When her brain was scanned to try to identify the cause, it was found that she had been born without a cerebellum. Although it had taken her some time to learn to walk as a child, and she was never able to run or jump, none of the medical staff who encountered her ever doubted that she was fully conscious. This was not a zombie, just a physically unstable human being.

If you examine the neural network at the heart of the cerebellum, you find patches that tend to be activated independently of one an-

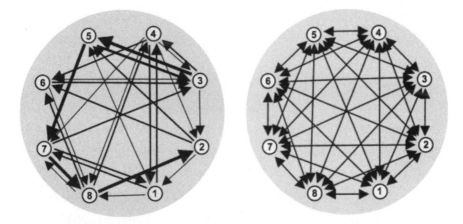

Left: A network with a high Φ and perhaps a higher level of consciousness.
Right: Although highly connected, the symmetry of this network causes little differentiation to occur across the network, resulting in the whole not contributing to the creation of new information not already inherent in the parts. This leads to a low Φ and hence a lower consciousness.

other with little interaction, just like the sleeping brain. The low Φ of the cerebellum's network fits in with the idea that it does not create conscious experience.

That the connectivity of the brain could be the key to consciousness has led to the idea that my "connectome," a comprehensive map of the neural connections in the brain, is part of the secret of what makes me "me." Just as the human genome project gave us unprecedented information about the workings of the body, mapping the human connectome may provide similar insights into the workings of the brain. Combine the wiring with the rules for how this network operates and we may have the ingredients for creating consciousness in a network.

Mapping the human brain is a distant goal, but already we have a complete picture of the wiring of the neurons inside *C. elegans*, a one-millimeter-long brainless worm with a fondness for compost heaps. Its nervous system contains exactly 302 neurons, ideally suited for us to map out a complete neuron-to-neuron wiring. Despite this map, we are still a long way from relating these connections to the behavior of *C. elegans*.

IS THE INTERNET CONSCIOUS?

Given that Tononi's Φ is a measure of how a network is connected, it could help us understand if my smartphone, the Internet, or even a city can achieve consciousness. Perhaps the Internet or a computer, once it hits a certain threshold at some point in the future, might recognize itself when it looks in the mirror. Consciousness could correspond to a phase change in this coefficient, just as water changes state when its temperature passes the threshold for boiling or freezing.

It is interesting that the introduction of an anesthetic into the body doesn't gradually turn down consciousness but at some point seems to suddenly switch it off. If you've ever had an operation and the anesthetist asks you to count to twenty, you'll know (or perhaps won't know) that there is a point at which you suddenly drop out. The change seems to be very nonlinear, like a phase transition.

If consciousness is about the connectivity of the network, what other networks might already be conscious? The total number of transistors that are connected via the Internet is of the order of 10^{18}, while the brain has about 10^{11} neurons. The difference lies in the way they are connected. A neuron is typically linked to tens of thousands of other neurons, which allows for a high degree of information integration. In contrast, transistors inside a computer are not highly connected. On Tononi's measure it is unlikely that the Internet is conscious . . . yet.

If I take a picture with the digital camera in my smartphone, the memory in my phone can store an extraordinary range of different images. A million pixels, for example, will give me $2^{1,000,000}$ different possible black-and-white images. It will record detail that I am completely unaware of when I look through the viewfinder. But that's the point—my conscious experience could not deal with that huge input. Raw sensory data are processed by the brain and integrated to glean the significant information and limit its content.

At present, computers are very bad at matching the power of human vision. A computer tends to read a picture pixel by pixel and finds it hard to integrate the information. Humans, however, are excellent at taking this huge amount of visual input and creating a story from the data. It is this ability of the human mind to integrate and pick out what is significant that is at the heart of Tononi's measure of consciousness Φ.

Of course, there is still no explanation of how or why a high Φ should produce conscious experience. We can certainly correlate it with consciousness, which is very suggestive—witness the experiments on sleeping patients—but we still can't say why this produces consciousness, or whether a computer wired in this way would experience consciousness.

There are reasons why having a high Φ might offer significant evolutionary advantages. It seems to allow one to plan for the future. By taking data from different sensors, the highly integrated nature of a network with a high Φ can make recommendations for future action. This ability to project oneself into the future, to perform mental time travel, seems to be one of the key evolutionary advantages of consciousness. A network with a high Φ appears to be able to demonstrate such behavior. But why couldn't it do all of this while being unconscious?

SKYPING CONSCIOUSNESS

Christof Koch, of the Jennifer Aniston neuron, is a big fan of Φ. I was keen to push him on whether he thought it was robust enough to provide an answer to the "hard problem," as philosopher David Chalmers famously put it, of getting inside another person's head.

Since I first met Koch at the top of Mount Baldy outside Caltech, he had taken up the directorship of an extraordinary venture called the Allen Institute for Brain Science, funded by Paul G. Allen, cofounder of Microsoft, to the tune of $500 million. Allen's motivation? He simply wants to understand how the brain works. It's all open science, it's not for profit, and his institute releases all of its data free. As Koch says, "It's a pretty cool model."

Since it was going to be tricky to make another trip to California to quiz Koch on his views on Φ, I decided the next best thing was to access Koch's consciousness via a call on Skype. But before that, Koch was keen to know which side of the philosophical divide I was on.

"Sure, we can chat about consciousness and the extent to which it is unanswerable. But I would hope that you're not going to turn away young people who are contemplating a career in neuroscience by buying in to the philosopher's conceit of the 'Hard problem' (hard with a capital H). Philosophers such as Dave Chalmers deal in belief systems and personal opinions, not in natural laws and facts. While they ask interesting questions and pose challenging dilemmas, they have an unimpressive historical record of prognostication."

Koch reminded me of Auguste Comte's prediction that we would never know what was at the heart of a star. It is a parable I have kept in mind ever since embarking on this journey into the unknown. He then added another important voice to the debate, that of his collaborator Francis Crick, who wrote in 1996, "It is very rash to say that things are beyond the scope of science."

Koch was in an ebullient mood when we finally connected via Skype—although to be honest, I've never seen him anything but excited about life at the cutting edge of one of today's greatest scientific challenges. Two days earlier, the Allen Institute had released data that tried to classify the different cell types in the brain.

"We've known for two hundred years that the body is made up of cells, but there are heart cells and skin cells and brain cells. But then we realized that there aren't just one or two types of brain cell. There may well be a thousand. Or several thousand. In the cortex proper, we just don't know."

Koch's latest research has produced a database detailing the different range of cortical neurons to be found in a mouse brain. "It's really cool stuff. But this isn't just a pretty picture in a publication as it usually is in science. It's all the raw data and all the data is downloadable." Koch believes that this kind of detailed scientific analysis of how the brain functions is key to chipping away at the challenge of understanding how the brain produces conscious experience.

I asked Koch what had inspired him to embark on this research project in the first place. At the time, many regarded setting off in search of consciousness as akin to heading into the Sahara desert in search of water.

"One reason was to show them all they were wrong," Koch said. "I like to provoke." Koch is also someone who likes taking risks. It's what drives him to hang off the side of a mountain, climbing unaided to the top. But there was an aspect to his choice that was more unexpected. It turned out that Koch's religious upbringing was partly responsible for his desire to understand consciousness.

"Ultimately, I think I wanted to justify to myself my continued belief in God. I wanted to show to my own satisfaction that to explain consciousness we need something more. We need something like my idea of a God. It turned out to be different. It provoked a great internal storm within me when I realized there wasn't really any need for a soul. There wasn't any real need for it to generate consciousness. It could all be done by a mechanism as long as we had something special like Tononi's integrated information theory, or something like that."

In some respects, Koch admits he failed. "Over the last ten years I lost my faith; I always tried to integrate it, but couldn't. I still very much have a sentiment that I do live in a mysterious universe. It's conducive to life. It's conducive to consciousness. It all seems very wonderful, and literally every day I wake up with this feeling that life is mysterious and a miracle. I still haven't lost that feeling."

Koch believes we are getting closer to demystifying consciousness, and he believes it is this mathematical coefficient Φ and Tononi's IIT that hold the key.

"I'm a big fan of this integrated information theory," he told me. "I have this mechanism, this brain, this computer currently in this one state. These neurons are off; these neurons are on; these transistors are switching; these are not. The theory says that, at least in principle, I can predict the experience of that system, not by looking at the input–output but by actually looking inside the system."

Koch has a tough battle on his hands if he hopes to convince others of his view. A hardcore group of philosophers and thinkers don't believe that science can ever really explain consciousness. They are members of a school of thought that goes by the name of mysterianism, which asserts that there are mysteries that will always be beyond the limits of the human mind. Top of the mysterians' hit list is the hard problem of consciousness.

The name of the movement—taken from a rock band called Question Mark and the Mysterians—was initially a term of abuse concocted by philosopher Owen Flanagan, who thought the movement's standpoint was extremely defeatist. As Flanagan wrote, "The new mysterianism is a postmodern position designed to drive a railroad spike through the heart of scientism." Koch has often found himself debating with mysterian philosophers who think he is misguided. So what does he say to those who question whether you can ever really know that a system is experiencing something?

"If you pursue that to its end you just end up with solipsism," he says. You can hear the exasperation in his voice. "Yes . . . solipsism is logically consistent, but then it's extremely improbable that everyone's brain is like mine, except only I have consciousness and you are all pretending. You can believe that, but it's not very likely."

Koch believes that seeking to pinpoint the brain states that correspond to certain experiences is as good as it's going to get if you want to explain consciousness. As he put it, "Well, Marcus, now you are seeing red or right now you are having the experience of smelling a rose. I can do fMRI and I can see that your rose neurons are activated, that your color area is activated. So in principle, this is an entirely empirical project."

Hardcore mysterians will contend that this does not explain *how* such a network gives rise to this sensation of feeling something. But aren't they in danger of demanding too much? We understand that the sensation of heat is just the movement of atoms. No one then suggests that we haven't really explained why the movement of atoms gives rise to heat. I think Koch would say the same of consciousness.

I wondered whether Koch thought we could ever get to a point at which we can use an fMRI scanner or something more sophisticated to know what someone is really experiencing.

"That's a pragmatic problem," he said. "That's the same problem that thermodynamics faced in the nineteenth century. You couldn't compute those things. But at least for a simple thing like a critter of ten neurons, the expressions are perfectly well defined and there is a unique answer. The maximum number of neurons one could navigate is a big practical problem. Will we ever overcome it? I don't know, because the problem scales exponentially with the number of neurons, which is really bad. But that's a different problem."

MAKING MINDS

The ultimate test of our understanding of what makes me "me" is to see whether we can build an artificial brain that has consciousness. Humans are actually quite good at making conscious things. My son is an example of how we can combine various cellular structures, an egg and sperm, and together they have all the kit and instructions for growing a new conscious organism. At least I have to assume that my son is conscious—on some mornings he appears to be at the zombie end of the spectrum.

The Human Brain Project, the brainchild of Henry Markram, aims within ten years to produce a working model of the human brain that can be uploaded onto a supercomputer. "It's going to be the Higgs boson of the brain, a Noah's archive of the mind," declared Markram. He was rewarded with a billion-euro grant from the European Commission to realize his vision. But when this is done, would these simulations be conscious? To my surprise, Koch didn't think so.

"Let's suppose that Henry Markram's brain project pans out and we have this perfect digital simulation of the human brain. It will also speak, of course, because it is a perfect simulation of the human brain, so it includes a simulation of Broca's area, responsible for language. The computer will not exist as a conscious entity. There may be a minimal amount of Φ at the level of a couple of individual transistors, but it will never feel like anything to be the brain."

Because it is a perfect simulation, it is going to object vociferously that it is conscious, that it has an internal world. But in Koch's view it would be all talk.

"If you get a computer to simulate a black hole, the simulated mass will not affect the real mass near the computer. Space will never be distorted around the computer simulation. In the same way, you can simulate consciousness on a computer, but the simulated entity will not experience anything. It's the difference between simulating and emulating. Consciousness arises from the cause–effect repertoire of the mechanism. It's something physical."

So it would seem that consciousness depends on how the network is put together. A similar issue is at work with the zombie network that Tononi constructed with ten neurons. The principle that Tononi used to turn a high-Φ network with ten neurons into a zero-Φ network with exactly the same input–output repertoire can be applied to the network of the human brain. Both would have the same input–output behavior, and yet, because of the difference in the internal state transitions, one has a high Φ, the other zero Φ. But how can we ever know whether this would mean that one is conscious and the other a zombie?

What bugs me is that this seems to highlight the near unanswerability of the question of consciousness. Φ is certainly a great measure of the difference between a simulation of the brain and the actual nature of the brain's architecture. It defines the difference between a zombie network and the human mind. But how can we really know that the zombie isn't conscious? It is still there shouting its head off: "For crying out loud, how many times do I need to tell you? I am conscious in here." But Φ says it's just pretending. I tell the zombie that Φ

says it doesn't have an internal world. But just as my conscious brain would, the zombie insists, "Yes I do."

One of the reasons Koch likes Φ as a measure of consciousness is that it plays into his panpsychic belief that we aren't the only ones with consciousness. If consciousness is about the integration of information across a network, then it can apply to things as small as an amoeba and as vast as the whole universe.

Consciousness is ultimately how much difference you make to yourself, the cause–effect power you have on yourself. So even if you take a minimal system like a cell—a single cell is already incredibly complex—it has some cause–effect power on itself. It will feel like something to be that cell. It will have a small but non-zero Φ. It may be vanishingly small, but it will have some level of consciousness. Of course, that's a very ancient intuition.

IS THERE CONSCIOUSNESS AFTER DEATH?

When I tell people I am investigating what we cannot know, a typical question is whether I'll be tackling life after death. This is intimately related to the question of consciousness. Does our consciousness live on? There is no good evidence that anything survives after death, but is there any way we can definitively know? Given the high level of correlation there is between brain activity and consciousness, it seems extremely unlikely that consciousness can survive beyond death. If some communication were possible, it might give us some hope. But no one outside Hollywood has yet succeeded in communicating with the dead.

When I put this question to Koch, he agreed with me and told me he had spent a week in India debating the issue with an unlikely collaborator: the Dalai Lama.

"It was very intense. Four hours in the morning, four hours in the afternoon, talking about science. Two whole days were dedicated to consciousness. Buddhists have been exploring their own conscious- ness from the inside using meditative techniques for two thousand years. In contrast, we do it using fMRI and electrodes and psychophys- ics from the outside, but basically our views tend to agree. He was very sympathetic to many of the scientific ideas and we agreed on many things. For example, he has no trouble with the idea of evolution."

But there must have been points of difference?

"The one thing where we fundamentally disagreed was this idea of reincarnation. I don't see how it would work. You have to have a mechanism that carries my consciousness or my memories over into the next life. Unless we find something in quantum physics, and I don't know enough about it to know, I don't see any evidence for it."

If you are going to attempt to argue for the possibility of reincar- nation as a scientist, you will need to come up with a plausible expla- nation of how consciousness survives the death of your body. There have been some interesting proposals. For example, if consciousness corresponds ultimately to patterns of information in the brain— something close to what Tononi is advocating—then some argue that information can theoretically always be reconstructed. Provided you avoid the black hole crematorium, the combination of quantum deter- minism and reversibility means no information is ever lost and thus some transfer might be possible.

The religious physicist John Polkinghorne offers a story for the possibility of life after death: "Though this pattern is dissolved at death, it seems perfectly rational to believe that it will be remembered by God and reconstituted in a divine act of resurrection." But nowa- days, not everyone resorts to God for the promise of resurrection. Some people are already exploring how to ensure that their brain's information is stored somewhere before their hardware packs up, but we are still decades away from this being possible. The idea of downloading my consciousness onto my smartphone so that "it" becomes "me" may not be so far from Polkinghorne's proposal. It's just Apple playing the role of God. "Rapture for nerds," as Koch calls it.

ZOMBIELAND

When faced with a question that we cannot answer, we have to make a choice. Perhaps the intellectually correct response is to remain agnostic. But what if your answer would have an impact on how you lead your life? Imagine if you regarded yourself as unique and everyone else as zombies—it would have a big effect on your interaction with the world.

Some posit that to crack the problem of consciousness we will have to add to our description of the universe a new basic ingredient that gives rise to consciousness, an ingredient that—like time, space, or gravity—can't be reduced to anything else. All we can do is explore the relationship between this new ingredient and the other ingredients we have to date.

To me this seems a cop-out. There are many examples in nature of critical tipping points where a phase transition happens, but rarely does the phase transition create something that can't be explained other than by creating a new fundamental entity. It's not without precedent, of course. Electromagnetic waves are created as a consequence of accelerating charged particles. That is not an emergent phenomenon, but a new fundamental ingredient of the natural world.

Others have suggested that consciousness may be associated with another state of matter, like solid or liquid. This state of matter has even been given a name: perceptronium, which is defined as the most general substance that feels subjectively self-aware. It sounds like a crazy idea, but perhaps, in order to get to grips with such a slippery concept, we need crazy ideas.

Some argue that nothing physical is ever going to answer the problem of what it feels like to be me, that the very existence of consciousness implies something that transcends our physical reality. But if mind can move matter, then something must connect these two realms that would allow science back into the game.

Perhaps the question of consciousness is not one of science, but one of language. This is the stance of a school of philosophy that traces its conception back to Ludwig Wittgenstein. In his *Philosophical Investigations*, Wittgenstein explores the problem of a private language. The process of learning the meaning of the word *table* involves people pointing at a table and saying, "Here is a table." If I point at something that doesn't correspond to what society means by "table," then I get corrected. Wittgenstein wonders whether the same principle can apply to more abstract concepts such as pain. Pain does not involve an object that is external to us, or that we can point at and call pain. We may be able to find a correlation within an output on an fMRI scanner that always corresponds to my feeling of pain. But when we point at the screen and say that is pain, what do we really mean by "pain"?

Wittgenstein believed that the problem of exploring the private worlds of our feelings was one of language. How can we share the meaning of "here is a pain"? It is impossible. "I have a pain in my tooth" seems to have the same quality as "I have a table," and it tempts us into assuming that there is something called "my pain" in the same sense that there is "a table." "The confusions which occupy us arise when language is like an engine idling," Wittgenstein explained. "Not when it is doing work."

If you have a feeling and declare, "I think this is pain," I have no way of correcting you. I have no criterion against which to determine whether what you are feeling is what I understand by the word *pain*. Wittgenstein explores whether it is possible to show you pain by pricking your finger and declaring, "That is pain." If you now reply,

"Oh, I know what 'pain' means; what I don't know is whether *this*, that I have now, is pain," Wittgenstein writes that "we should merely shake our heads and be forced to regard [your] words as a queer reaction which we have no idea what to do with."

We can get at the problem of identifying what is going on inside our heads by imagining that we each have a box with something inside. We all name the thing inside "a beetle," but we aren't allowed to look in anyone else's box, only our own. So it is quite possible that everyone has something different in the box, or that the thing inside is constantly changing, or even that there is nothing in the box. And yet we all call it "a beetle." But for those with nothing in the box, the word *beetle* is a name for nothing. Does this word therefore have any meaning? Are our brains like the box and consciousness the beetle? Does an fMRI scanner allow us finally to look inside someone else's box and see if your beetle is the same as mine? Does it rescue consciousness from Wittgenstein's language games?

Wittgenstein explores how a sentence can fool us into thinking it means something because it takes exactly the form of a real sentence, but when you examine it carefully you find that it doesn't actually refer to anything. This, many philosophers believe, is at the heart of the problem of consciousness. It is not a question of science, but simply a confusion of language.

Daniel Dennett is one of the most famous of the modern-day philosophers who follow in the tradition of Wittgenstein. He believes that in years to come we will recognize that there is little point in having this long debate about the concept of consciousness. To his mind, if we do encounter a machine or organism that functions exactly as we would expect a conscious being to, we should simply define this thing as conscious. The question of whether it is a zombie, whether it is actually feeling anything, should just be ignored. If there is no way to distinguish the unconscious zombie from the conscious human, what point is there in having a word to describe this difference? Only when you pinpoint the difference is there any point in having a word to describe it.

Dennett cites vitalism, the idea that there is some extra special ingredient, an *elan vital*, that breathes life into a collection of cells to support his stance. If you show me a collection of cells that have the ability to replicate themselves, that meows and purrs like a cat, and then declare that it isn't actually alive, you would have a very hard time convincing me that you are correct. Some believe that arguments over consciousness will go the same way as those over vitalism, which came into being as an attempt to explain how a physical system could do the things associated with a living organism: reproduce, self-organize, adapt. Once the mechanisms were explained, we had solved the problem of life and no *elan vital* was needed. But in the case of consciousness, some believe that the question of how physical stuff produces a subjective experience cannot by its very nature be answered mechanistically.

With the advent of extraordinary new equipment and techniques, our internal world has become less opaque. Mental activities are private, but their physical manifestations are public. To what extent can I gain access to private thoughts by analyzing their traces? We can probably get to the stage where I will know that you are thinking about Jennifer Aniston rather than Pythagoras's theorem just by looking at the neurons that are firing. But to get a sense of what it feels like to be you—is that possible just by understanding the firing of neurons? Will I ever be able to distinguish the conscious from the zombie?

My gut has as many as a hundred million neurons, 0.1 percent of the neurons in my brain, yet it isn't conscious. Or is it? How can I be certain that it doesn't actually have a distinct consciousness? My stomach might start communicating with me, but does that communication imply it is conscious? I might scan it, probe it with electrodes, discover that the wiring and firing are a match for a cat brain that has as many neurons, but is that as far as we can go?

The question of how we can definitively distinguish the zombies from the conscious may be one of the great unknowns—the unanswerable questions of science. The question that I put to my smartphone at the beginning of this Edge hints at the challenge ahead. Was it the zombie chatbot who wanted to become a poet, or did it want to become rich? Is it clever enough to make a joke about Descartes's "I think, therefore I am?" Will it eventually start dating? Who is the zombie and who is conscious?

As we finished our Skype chat, Koch conceded that there was no guarantee we would ever know how to determine consciousness.

"There is no law in the universe which says we have the cognitive power to explain everything. If we were dogs—I mean, I love my dog—my dog is fully conscious, but my dog doesn't understand taxes, it doesn't understand special relativity, or even the simplest differential equation. I really dislike it when people say you will never know this. Yes, there's no guarantee. But it's a really defeatist attitude, right? I mean, what sort of research program is it, Marcus, where you throw up your hands and say forget about it, I can't understand it ever, it's hopeless? That's defeatism."

With that clarion call not to give up trying to answer life's unanswerable questions we ended our Skype call. Koch's face vanished from my screen, leaving me with a slightly uneasy feeling. Was I sure it had been Koch at the other end of the line? Or had he devised some algorithmically generated avatar to handle my onslaught of inquiries? Our conversation had presented so many tantalizing possibilities.

SEVENTH EDGE

INFINITY

13

Number is the ruler of forms and ideas and the cause of gods and demons.

—Pythagoras

CHRISTMAS DINNER IN ENGLAND consists of turkey, a flaming Christmas pudding, and Christmas crackers, colorfully decorated tubes that explode with a bang as two diners pull each end to reveal a toy, a joke, and a tissue paper crown that is then worn throughout the dinner. Bored of the uninspiring range of Christmas crackers available in the shops, I decided this year to treat my family to my very own homemade mathematical crackers. Each one included a math joke and a paradox. My children were of the unanimous opinion that the jokes were stronger on the mathematics than the humor. I'll let you decide . . . How many mathematicians does it take to change a lightbulb? 0.99999 recurring. And if you're not laughing, don't worry. My family wasn't either. If you don't get it, then—although you should never have to explain a joke—the point is that you can prove that 0.9999 recurring is actually the same number as 1.

The paradoxes were a little less lame. One contained a Möbius strip, a seemingly paradoxical geometric object that has only one side. If you take a long strip of paper and twist it before gluing the ends together, the resulting shape has only one side. You can check this by trying to color the sides: start coloring one side and you soon find you've colored the whole thing. The strip has the surprising property that if you cut it down the middle, it doesn't come apart into two loops, as you might expect, but remains intact. It's still a single loop but with two twists in it.

The cracker I ended up with wasn't too bad, if I say so myself. The joke was quite funny. What does the B in Benoit B. Mandelbrot stand for? Benoit B. Mandelbrot. If you still aren't laughing, the thing you're missing is that Mandelbrot discovered the fractals that were featured in my First Edge, those geometric shapes that never get simpler, however much you zoom in on them. And the paradox was one of my all-time favorites. It consisted of two statements, one on either side of a card, as follows:

The statement on the other side of this card is false.
The statement on the other side of this card is true.

I've always enjoyed and in equal measure been disturbed by word games of this sort. One of my favorite books as a kid was called *What Is the Name of This Book?* It was stuffed full of crazy word games that often exploited the implications of self-reference.

I have learned not to be surprised by sentences that give rise to paradoxes like the one captured by the circular logic of the two sentences on the card in my Christmas cracker. Just because you can form a meaningful sentence doesn't mean it is true or false. I think the slippery nature of language is one of the reasons that I was drawn toward the certainties of mathematics, where this sort of ambiguity was not tolerated. But, as we shall discover in this Edge, my cracker paradox was used by one of the greatest mathematical logicians of all time, Kurt Gödel, to prove that even my own subject contains true statements about numbers that we will never be able to prove are true.

SCIENCE VERSUS MATH

The desire for certainty, to know—to *really* know—was one of the principal reasons why I chose mathematics over the other sciences. In science, the things we think we know about the universe are models that match the data. To qualify as scientific, the model must be capable of potentially being falsified or proved wrong. A theory can only survive if all evidence supports it. As soon as evidence is discovered that contradicts the model, we must change our theory. By its very nature, a scientific theory is thus something that can potentially be thrown out. So how can we ever truly know we've got it right?

We thought the universe was static, but then new discoveries revealed that galaxies were racing away from us. We thought the universe was expanding at a rate that was slowing down, given the drag of gravity, but then we discovered that the expansion was accelerating. We modeled this with the idea of dark energy pushing the universe apart, and that model waits to be falsified. Eventually, we may well hit on the right model of the universe, but we'll never know for sure.

One of the exciting things about science is that it is constantly evolving—there are always new stories. I sometimes feel rather sorry for the old stories that fade into irrelevance. Of course, the new stories grow out of the old. As a scientist you live with the fear that your theory may be the flavor of the moment, winning prizes only to be suddenly superseded, like the plum pudding model of the atom, or the idea of absolute time. These are no longer atop the science bestseller list. They have been replaced by new stories.

The model of the universe I read about as a schoolkid has been completely rewritten. The same cannot be said of the mathematical theorems I learned. They are as true now as they were the day I first read them and the day they were first discovered, sometimes as long as two thousand years ago. As an insecure, spotty teenager, I found that certainty attractive. That's not to say that mathematics is static. It is constantly growing as unknowns become known; but the knowns remain known and robust, and form the first pages in the next great story. Why is the process of attaining mathematical truth so different from that faced by the scientist who can never really know?

The all-important ingredient in the mathematician's cupboard is the proof.

PROOF: THE PATH TO TRUTH

Archeologists have found evidence of people doing math as far back as the second millennium BC. Clay tablets in Babylon and papyri in Egypt show sophisticated calculations and puzzles being solved: estimates for π, the formula for the volume of a pyramid, algorithms to solve quadratic equations. But in general, these documents describe procedures that can be applied to particular problems to derive solutions. We don't find justification for why these procedures should always work, beyond the evidence that it's worked in the thousands of examples that have been documented in the clay tablets to date. Mathematical knowledge was originally based on experience and had a more scientific flavor to it. Procedures were adapted if a problem cropped up that wasn't amenable to the current algorithm.

Then, around the fifth century BC, things began to change as the ancient Greeks got their teeth into the subject. The algorithms started to come with arguments to justify why they would always do what it says on the tin or tablet. It isn't simply that it's worked the last thousand times, so it will probably work the next time too—the argument explains why the proposal will always work. The idea of proof was born.

Thales of Miletus is credited with being the first known author of a mathematical proof. He proved that if you take any point on the circumference of a circle and join that point to the two ends of a diagonal across the circle, then the angle you've created is an exact right angle. It doesn't matter which circle you choose or what point on the circle you take, the angle is always a right angle. Not approximately, and not because it seems to work for all the examples you've drawn, but because it is a consequence of the properties of circles and lines.

Thales's proof takes a reader from things they are confident are true and, by a clever series of logical moves, arrives at this new point of knowledge, which is not one that you would necessarily think obvious just from looking at a circle. The trick is to draw a third line going from the initial point B on the circle to the center of the circle at O.

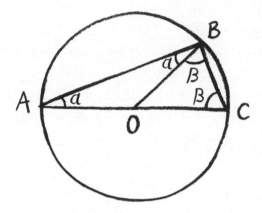

Why does this help? Now you have two triangles with two sides of equal length. This means that in both triangles the two angles opposite the center of the circle are equal. This is something that has already been proved about such triangles. Take the larger triangle you originally drew. Its angles add up to $2\alpha + 2\beta$. Combine this with the knowledge that a triangle has angles adding up to 180 degrees, and we know that $\alpha + \beta$ must be 90 degrees, just as Thales asserted.

When I first saw this proof as a kid it gave me a real thrill. I could see from the pictures that the angle on the circle's edge looked like a right angle. But how could I know for sure? My mind searched for some reason why it must be true. And then as I turned the page and saw the third line Thales drew to the center of the circle, and took in the logical implications that flowed from that, I suddenly understood with thunderous clarity why that angle must indeed be 90 degrees.

Note that already in this proof you see how the mathematical edifice is raised on top of things that have already been proved, things like the angles in a triangle adding up to 180 degrees. Thales's discovery in turn becomes a new block on which to build the next layer of the mathematical edifice.

Thales's proof is one of the many to appear in Euclid's *Elements*, the book many regard as the template for what mathematics and proof are all about. The book begins with a few basic building blocks or axioms, statements of geometry that seem so blindingly obvious that you are prepared to accept them as secure foundations on which to start building logical arguments.

The idea of proof wasn't created in isolation. It emerged alongside a new style of writing that developed in ancient Greece. The art of rhetoric, formulated by the likes of Aristotle, provided a new form of discourse aimed at persuading an audience. Whether in a legal context or political setting, or simply the narrative arc of a story, audiences were taken on a logical journey as the speaker attempted to convince listeners of his position. Just as mathematics emerged in Egypt and Babylon to support the building and measuring of new cities growing up around the Euphrates and Nile, logic and rhetorical argument emerged from the political institutions of Greece's flourishing city-states.

Aristotle defined rhetoric as a combination of pure logic and strategic emotional manipulation. Mathematical proofs are more logic than emotion, but a proof is also about storytelling. The development of the proof at this time and place probably had as much to do with the sophisticated narratives of Sophocles and Euripides as with the philosophical dialogues of Aristotle and Plato. The mathematical explorations of the Greeks moved beyond functional algorithms for building and surveying to surprising and exciting discoveries.

A proof is a logical story that takes a reader from a place they know to a new destination. Like Frodo's adventures in J.R.R. Tolkien's *Lord of the Rings*, it describes the journey from the Shire to Mordor. Within the boundaries of the familiar land of the Shire are our well-known axioms, the self-evident truths about numbers, together with those propositions that have already been proved. This is the setting for the beginning of the quest. The rules of mathematical deduction, like the legitimate moves of a chess piece, establish the steps you are permitted to take through this world. At times you arrive at what looks like an impasse and will need to take a lateral step, moving sideways or even backward to find a way around. Sometimes you need to wait centuries for new mathematical characters like imaginary numbers or calculus to be created so you can continue on your journey. The proof is the map charting the coordinates of your journey. It is the mathematician's log.

To earn its place in the mathematical canon it isn't enough that the journey produces a true statement about numbers or geometry. It needs to surprise, delight, move the reader. It should contain drama and jeopardy. Mathematics is distinct from the collection of all true statements about numbers, just as literature is not the set of all possible combinations of words or music all possible combinations of notes. Mathematics involves aesthetic judgment and choice. This is probably why the art of the mathematical proof grew out of a period when the art of storytelling was flourishing. Proof probably owes as much to the pathos of Aristotle's rhetoric as to its logos.

NUMBERS AT THE EDGE

While many of the first geometric proofs are constructive, the ancient Greeks also used their new mathematical tools to prove that certain things are impossible, that they lay beyond knowledge. The discovery, as we saw, that the square root of 2 cannot be written as a fraction is one such striking example.

The proof has a narrative quality to it, and at first one assumes that we will reach our hypothetical destination of expressing the square root of 2 as a fraction. As the story innocently unfolds, one gets sucked further and further down the rabbit hole until finally a completely absurd conclusion is reached: odd numbers are even and vice versa. The moral of the tale is that the imaginary fraction representing the square root of 2 must be an illusion. (If you would like to take a trip down the rabbit hole, the story is reproduced in the box below.

For those encountering a number like the square root of 2 for the first time, it must have seemed like something that by its nature was incomprehensible. To know a number was to write it down, to express it in terms of other numbers you knew. But here was a number that seemed to defy any attempt to record its value.

Proof of the Irrationality of the Square Root of 2

Let L be the length of the hypotenuse of a right-angled triangle whose short sides are both of length 1 unit. Pythagoras's theorem implies that a square placed on the hypotenuse will have the same area as the sum of the area of squares placed on the two smaller sides. But these smaller squares each have area 1 while the larger square has area L^2. Therefore L is a number which, when squared, gives you 2.

Suppose L is actually equal to a fraction $L = p/q$.

We can assume that one of p or q is odd. If they are both even, we can keep dividing both top and bottom by 2 until one of the numbers becomes odd.

Since $L^2 = 2$, it follows that $p^2/q^2 = 2$.

Multiply both sides by q^2: $p^2 = 2 \times q^2$.

So is p odd or even? Well p^2 is even, so p must be even because odd times odd is still odd. So $p = 2 \times n$ for some number n. Since p is even, that means that q must be the odd number. But hold on . . .

$2 \times q^2 = p^2 = (2 \times n)^2 = 2 \times 2 \times n^2$,

so we can divide both sides of the equation by 2 to get

$q^2 = 2 \times n^2$.

Remember that we'd worked out that q was the odd number.

So q^2 is also odd. But the right-hand side of this equation is an even number! So if L is expressible as a fraction it would imply that odd = even. That's clearly absurd, so our original assumption that L can be written as a fraction must have been false.

I think this is one of the most stunning proofs in mathematics. With just a finite piece of logical argument we have shown there is a length that needs infinity to express it.

It was an extraordinary moment in the history of mathematics: the creation of a genuinely new sort of number. You could have taken the position that the equation $x^2 = 2$ doesn't have any solutions. At the time, no known number could solve this equation precisely. Indeed, it was really only in the nineteenth century that sufficiently sophisticated mathematics was developed to make sense of such a number. And yet you felt it did exist. You could see it—there it was, the length of the side of a triangle. Eventually, mathematicians took the bold step of adding new sorts of numbers to our mathematical tool kit so that we could solve these equations.

There have been other equations that seemed to defy solution, where the answer was not as visible as the square root of 2, and yet we've managed to create solutions. Solving the equation $x + 3 = 1$ looks easy from our modern perspective: $x = -2$. But the Greeks had no conception of negative numbers. Diophantus refers to this sort of equation as absurd. For Greek mathematicians like Diophantus, numbers were geometrical: they were real things that existed, lengths of lines. There was no line such that if you extended it by three units it would be one unit in length.

Other cultures were not so easily defeated by such an equation. In ancient China numbers counted money, and money often involved debt. There certainly could be a circumstance in which I could add three coins to my purse, only to find I had one coin left. A debt to a friend could be responsible for swallowing two of the coins. In 200 BC Chinese mathematicians depicted their numbers using red sticks; but if they represented debts, they used black sticks. This is the origin of the term "going into the red," although somewhere along the line for some reason the colors were swapped.

It was the Indians, in the seventh century AD, who first success-fully developed a theory of negative numbers. Brahmagupta even de-duced some of the important mathematical properties of these numbers—for example, that "a debt multiplied by a debt is a fortune," or minus times minus is plus. Interestingly, this isn't a rule; it's a con-sequence of an axiom. It is a fun challenge to prove why this must be so. It took Europeans until the fifteenth century to be convinced of the existence of negative numbers. Their use was at first so disruptive that they were banned in thirteenth-century Florence.

Over the centuries new numbers continued to appear, as mathe-maticians were faced with the challenge of solving an equation like $x^2 = -1$. At first sight, it seems impossible. After all, if I take a positive number and square it, the answer is positive, and, as Brahmagupta proved, a negative number squared is also positive. When mathema-ticians in the Renaissance encountered this equation, their first reac-tion was to assume it was impossible to solve. Then Italian mathematician Rafael Bombelli took the radical step of supposing there was a new number that had a square equal to -1, and he found that he could use this new number to solve a whole host of equations that had been thought unsolvable. Interestingly, this mysterious number would sometimes be needed only in the intermediate calculations and would disappear from the final answer, leaving ordinary numbers that people were familiar with that clearly solved the equation at hand.

It was like mathematical alchemy, but many refused to admit Bombelli's new number into the canon of mathematics. Descartes wrote about them in a rather derogatory manner, dismissing them as imaginary numbers. As time went on, more mathematicians realized their power and appreciated the fact that they didn't seem to give rise to any contradictions if they were admitted. It wasn't until the beginning of the nineteenth century that imaginary numbers truly found their place in mathematics, thanks in part to a picture that helped mathematicians to visualize them.

Ordinary numbers (or what mathematicians now call *real* numbers) were laid out along the number line running horizontally across the page. The imaginary numbers like i, the name given to the square

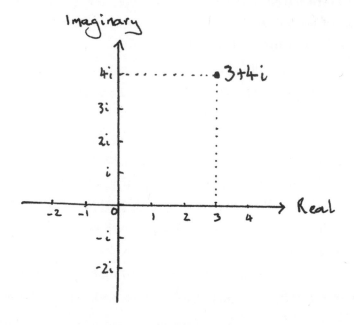

root of –1, and multiples of *i* were pictured on the vertical axis. This two-dimensional picture of the imaginary or complex numbers was instrumental in their acceptance. The potency of this picture was demonstrated by the discovery that its geometry reflected the arithmetic of the numbers.

So are there even more numbers out there beyond the edge that we haven't yet discovered? I might try writing down more strange equations in the hope that I get more new numbers. For example, what about the equation $x^4 = -1$?

Perhaps to solve this equation I'll need even more new numbers. But one of the great theorems of the nineteenth century, now called the Fundamental Theorem of Algebra, proved that using the imaginary number *i* and real numbers, I can solve any equation built out of powers of *x* that I want. For example, if I take

$$x = \frac{1}{\sqrt{2}} + \frac{1}{\sqrt{2}} i$$

and raise it to the power of 4, then the answer is –1. The edge had been reached beyond which no new numbers would be found by trying to solve equations.

PROVING THE IMPOSSIBLE

The ancient Greek proof of the irrationality of the square root of 2 was the first of many mathematical proofs to show that certain things are impossible. Another proof of the impossible was the concept of "squaring the circle," a concept that has entered many languages as an expression of impossibility. Squaring the circle was a geometric challenge that the ancient Greeks enjoyed trying to crack, using a ruler (without measurements) and a compass (to make the arcs of circles). They discovered clever ways to use these tools to draw perfect equilateral triangles, pentagons, and hexagons, but never managed to square a circle.

Squaring the circle involves trying to construct from any given circle a square whose area is the same as that of the circle you started with. Try as hard as the Greeks could, the problem stumped them.

A similarly impossible challenge was posed by the oracle of the island of Delos. Residents of Delos had consulted the oracle to get advice about how to defeat a plague that the god Apollo had sent down upon them. The oracle replied that they must double the size of the altar to Apollo. The altar was a perfect cube. They could have just put two copies of the cube side by side to make a rectangular box, but that was too easy. Plato interpreted this as a challenge by the oracle to construct, using straight edge and compass, a second perfect cube whose volume was double the volume of the first cube.

If the second cube is double the volume of the first, this means the sides have lengths that are the cube root of twice the length of the first cube. The challenge, therefore, was to construct a length equal to the cube root of 2. Constructing a length that is the square root of 2 was easy, since it is the diagonal of a unit square; but the cube root of 2 turned out to be so difficult that the residents of Delos could not solve the problem. Perhaps the oracle wanted to distract the Delians with geometry to divert their attention from the more pressing social problems they faced as a result of the plague.

Doubling the cube, squaring the circle, and a third classical problem, trisecting the angle, all turned out to be impossible. But it wasn't until the nineteenth century that mathematicians could prove beyond a doubt that this was the case. The development of group theory by the French revolutionary Evariste Galois in the 1820s—a language for understanding symmetry that I use in my own research—held the key to proving the impossibility of these geometric constructions. It turns out that the only lengths that can be constructed by a ruler and compass are those that are solutions to certain types of algebraic equation.

In the case of squaring the circle, the challenge requires you to produce, using a ruler and compass, a straight line of length $\sqrt{\pi}$. But it was proved in 1882 that π is not only an irrational number but also a transcendental number, which means it isn't the solution to any algebraic equation. This in turn means that it is impossible to square the circle.

Mathematics is remarkably good at proving when things are impossible. One of the most celebrated theorems on the mathematical books is Fermat's Last Theorem, which states that it is impossible to find any whole numbers that will solve any of the equations $x^n + y^n = z^n$, where n is greater than 2. By contrast, $n = 2$ is easy, as it corresponds to the equation Pythagoras produced from his right-angled triangles. If $n = 2$, there are many solutions, for example, $3^2 + 4^2 = 5^2$. In fact, there are infinitely many solutions, and the ancient Greeks found a formula to produce them all. But it is often much easier to find a solution than to prove that you will never find one.

Fermat famously thought he had found an argument that proved that his equations had no solutions. He scribbled in the margin of his copy of Diophantus's *Arithmetica* that the margin was too small for his remarkable proof. Then he died without ever elaborating. It took 350 years before my Oxford colleague Andrew Wiles finally produced a convincing argument to explain why you will never find whole numbers that solve Fermat's equations. Wiles's proof runs to more than one hundred pages, in addition to the thousands of pages of preceding theory that it is built on. So even a very wide margin wouldn't have sufficed.

The proof of Fermat's Last Theorem is a tour de force. I regard it as a privilege to know the man who cracked it and to have been alive when the final pieces of the proof were put in place.

Before Wiles had shown the impossibility of finding a solution, there was always the prospect that maybe there were some sneaky numbers out there that did solve one of these equations. I remember a great April Fool's joke that went around the mathematical community around the same time as Wiles announced his proof. The punchline was that a respected number theorist at Harvard, Noam Elkies, had produced a nonconstructive proof that solved Fermat's equations. It was a cleverly worded April Fool's e-mail because the "nonconstructive" bit meant he couldn't explicitly say what the numbers were, just that his proof implied there must be a solution. Most people were forwarded the e-mail several days after the joke first started its rounds, and so had no clue it was a joke!

April Fool's or not, the mathematical community had spent 350 years not knowing whether there was a solution. But finally Wiles put us out of our misery. His proof means that, however many numbers I try, I'll never find three that make Fermat's equations work.

RUNNING OUT OF NEURONS

We are living in a golden age of mathematics, a time when some of the great unsolved problems have finally been cracked. In 2003, Russian mathematician Grigori Perelman proved the Poincaré conjecture, one of the great challenges of geometry. But there are still many statements about numbers and equations whose proofs elude us: the Riemann hypothesis, the twin primes conjecture, the Birch–Swinnerton-Dyer conjecture, Goldbach's conjecture.

In my own research I have dedicated the last twenty years to trying to settle whether something called the PORC conjecture is true or not. The PORC conjecture was formulated more than fifty years ago by Oxford mathematician Graham Higman, who believed that the number of symmetry groups with a given number of symmetries should be given by nice polynomial equations (this is what the P in PORC stands for). For example, the number of symmetry groups with p^6 symmetries, where p is a prime number, is given by a quadratic expression $p^2 + 39p + c$ (where c is a constant that depends on the remainder you get on dividing p by 60).

My own research has cast real doubts on whether this conjecture is true. I discovered a symmetrical object with p^9 symmetries whose behavior suggests something very different to what was predicted by Higman's conjecture. But it doesn't completely solve the problem. There is still a chance that other symmetrical objects with p^9 symmetries could smooth out the strange behavior I unearthed, leaving Higman's conjecture true. So at the moment, I don't know whether this conjecture is true, and Higman, alas, died before he could know. I am desperate to know before my finite time on earth comes to an end, and it is questions like this that drive my mathematical research.

I sometimes wonder, as I get lost in the seemingly infinite twists and turns of my research, whether my brain will ever have the capacity to crack the problem I am working on. In fact, mathematics can be used to prove that there are mathematical challenges beyond the physical limitations of the human brain, with its eighty-six billion neurons connected by more than a hundred trillion synapses.

Mathematics is infinite. It goes on forever. Unlike chess, which has an estimated $10^{10^{50}}$ possible different games, the provable statements of mathematics are infinite in number. In chess, pieces are taken, games are won, and sequences are repeated. Mathematics, though, doesn't have an endgame, which means that even if I get my eighty-six billion neurons firing as fast as they physically can, there are only a finite number of logical steps that I will be able to make in my lifetime, and therefore only a finite amount of mathematics that I can ever know. What if my PORC conjecture needs more logical steps to prove than I will ever manage in my lifetime?

Even if we turned the whole universe into a big computer, there are limits to what it could ever know. In his paper "Computational

Capacity of the Universe," Seth Lloyd calculates that since the Big Bang the universe cannot have performed more than 10^{120} operations on a maximum of 10^{90} bits. This means that at any point in time, the universe can only know a finite amount of mathematics. You might ask, what is the universe calculating? It's actually computing its own dynamical evolution. Although this is a huge number, it is still finite. It means that we can prove computationally that at any point in time, there will always be things that the observable universe will not know.

But it turns out that mathematics has an even deeper layer of unknown. Even if we had a computer with infinite capacity and speed, there are things it would never know. A theorem proved in 1930 has raised the horrific possibility that even a computer with this infinite capacity may never know whether my PORC conjecture is true. Called Gödel's incompleteness theorem, it shook mathematics to the core. Gödel proved that within any axiomatic framework for mathematics there are mathematically true statements that we will never be able to prove are true within that framework. A mathematical proof that there are things beyond proof—mathematics beyond the edge.

When I learned about this theorem at university, it had a profound effect on me. Despite my own physical limitations, I grew up with the comforting belief that, at least theoretically, there was a proof out there that would demonstrate whether the Riemann hypothesis was true or not. One of my mathematical heroes, the Hungarian Paul Erdős, used to speak affectionately about proofs from "The Book." This was where Erdős believed God kept the most elegant proofs of each mathematical theorem. The challenge for a mathematician is to uncover the proofs in The Book. As Erdős joked in a lecture he gave in 1985, "You don't have to believe in God, but you should believe in The Book." Erdős himself doubted the existence of God, dubbing him the Supreme Fascist and holding him responsible for hiding his socks and passport. But I think most mathematicians accept the metaphor of The Book. The trouble was, the proof of Gödel's incompleteness model, which I learned about in my university course on mathematical logic, proved that there were pages missing from The Book, pages that even the Supreme Fascist did not possess.

PARALLEL UNIVERSES

The revelation that there were mathematical statements beyond the limits of proof was sparked by the realization that one of Euclid's axioms was not as axiomatic as had been thought.

An axiom is a premise or starting point for any logical train of thought. In general, an axiom is meant to express a self-evident truth that is accepted as true without the need for proof. For example, I believe that if I take two numbers and add them together, I will get the same answer regardless of the order of the equation. If I take 36 and add 43, I won't get a different answer than if I start with 43 and add 36. You might ask how I know that this will always be true. Perhaps something weird happens if I take really large numbers and add them together. What mathematics does is to make deductions from axioms. If the way we count things in the universe does something weird, we just have to accept that the mathematics that we have developed from this axiom is not applicable to the way physical numbers work in the universe. We would then develop a new number theory based on numbers that satisfy a different fundamental set of axioms.

Although most of the axioms that Euclid used to deduce his theory of geometry seemed self-evident, there was one that mathematicians increasingly thought a bit suspicious.

The parallel postulate states that if you have a line and a point off that line, then there is a unique line through the point that is parallel to the first line. This postulate certainly seems logical if you are drawing geometry on a flat page. It is one of the axioms that is crucial to Euclid's proof that triangles have angles that add up to 180 degrees. Certainly any geometry in which the parallel postulate is true will give rise to triangles with this property. But the discovery in the nineteenth century of new sorts of geometry with no parallel lines, or with many lines that could be drawn parallel, led mathematicians to the realization that Euclid's geometry was just one of a whole host of different possible geometries.

If I take the surface of a sphere—a curved geometry—then lines confined to this surface are not straight, but bend. Take two points on the surface of the Earth for example, and as anyone who has flown across the Atlantic knows, the shortest path between these two points won't match the line you would draw on a flat atlas. This is because the line between these two points is part of a circle, like a line of longitude that divides the sphere perfectly in half. Indeed, if one of the points is a North or South Pole, then the line would be a longitudinal line. All lines in this geometry are just longitudinal lines that are moved around the surface so that they pass through the two points you are interested in. They are called great circles. But now if I take a third point off this great circle, there is no way to arrange another great circle through the third point that won't intersect the first. So here is a geometry that doesn't have parallel lines.

Consequently, any proof that depends on the parallel postulate won't necessarily be true in this new geometry. Take the proof that angles in a triangle add up to 180 degrees. This statement depends on the geometry satisfying the parallel postulate. But this curved geometry doesn't. And indeed, this geometry contains triangles whose angles add up to more than 180 degrees. Take the North Pole and two points on the equator and form a triangle through these three points. Already the two angles at the equator total 180 degrees, so the total of all three angles will be greater than 180 degrees.

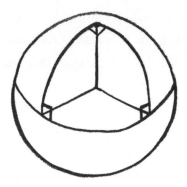

A triangle whose angles add up to more than 180 degrees

Other geometries were discovered in which there was not just one but many parallel lines through a point. In these geometries, called hyperbolic, triangles have angles that add up to less than 180 degrees. These discoveries didn't invalidate any of Euclid's proofs; mathematical breakthroughs only enrich rather than destroy previous knowledge. But the introduction of these new geometries in the early nineteenth century nonetheless caused something of a stir. Indeed, some mathematicians felt that a geometry that didn't satisfy Euclid's axiom about parallel lines must contain some contradiction that would lead to that geometry being thrown out. But further investigation revealed that any contradiction inherent in these new geometries would imply a contradiction at the heart of Euclidean geometry.

Such a thought was heresy. Euclidean geometry had not produced a contradiction for two thousand years. But hold on . . . this is beginning to sound like the way we do science! Surely mathematics should be able to prove that Euclidean geometry won't give rise to contradictions. We shouldn't just rely on the fact that it's worked up to this point. That's how the scientists in the labs across the road work. We mathematicians should be able to prove that our subject is free of contradictions.

WHAT IS THE NAME OF THIS SECTION?

When, at the end of the nineteenth century, mathematical set theory produced strange results that seemingly led to unresolvable paradoxes, mathematicians began to take more seriously this need to prove our subject as free of contradictions. The British philosopher Bertrand Russell came up with a number of these worrying paradoxes. He challenged the mathematical community with the set of all sets that don't contain themselves as members. Was this new set a member of itself? The only way you could be in this new set was if you were a set that didn't contain you as a member. But as soon as this set entered the new set, suddenly it is (of course) a set that contains itself as a member. Arrgh! There seemed no way to resolve this paradox.

Russell came up with a few more prosaic examples of this sort of self-referential paradox. For example, he imagined an island on which a law decreed that the barber had to shave all those people who didn't shave themselves—and no one else. The trouble is that this law leads to a paradox—can the barber shave himself? No, since he is allowed only to shave those who don't shave themselves. But then he qualifies as someone who must be shaved by the barber. Arrgh, again! The barber has taken the role of the set Russell was trying to define, the set of all those sets that don't contain themselves as members.

I think my favorite example of all these paradoxes is the puzzle of describing numbers. Suppose you consider all the numbers that can be defined using fewer than twenty words taken from the *Oxford English Dictionary*. For example, 1,729 can be defined as "the smallest number that can be written as the sum of two cubes in two different ways." Since there are only finitely many words in the *Oxford English Dictionary* and we can use only a maximum of twenty words, we can't define every number in this way, as there are infinitely many numbers and only finitely many sentences with fewer than twenty words. So there is a number defined as follows: "the smallest number that cannot be defined in fewer than twenty words from the *Oxford English Dictionary*." But hold on . . . I just defined it in fewer than twenty words. Arrgh!

Natural language is prone to throw up paradoxical statements. Just putting words together doesn't mean they make sense or have a truth value. But the worrying aspect to Russell's set of all sets that don't contain themselves is that it is very close to the sorts of things we might want to define mathematically.

Mathematicians eventually found a way around this paradoxical situation, which required refining the intuitive idea of a set, but it still left a worrying taste in the mouth. How many other surprises were in store? When the great German mathematician David Hilbert was asked to address the International Congress of Mathematicians in 1900, he decided to set out the twenty-three greatest unsolved problems facing mathematicians in the twentieth century. Proving that mathematics was free of contradictions was second on his list.

In his speech, Hilbert boldly declared what many believed to be the mantra of mathematics: "This conviction of the solvability of every mathematical problem is a powerful incentive to the worker. We hear within us the perpetual call. There is the problem. Seek its solution. You can find it by pure reason, for in mathematics there is no *ignorabimus*." In mathematics there is nothing that we cannot know. A bold statement indeed.

Hilbert was reacting to a growing movement at the end of the nineteenth century that held that there were limits to our ability to understand the universe. The distinguished physiologist Emil du Bois-Reymond had addressed the Berlin Academy in 1880, outlining what he regarded as seven riddles about nature that he believed were beyond knowledge, declaring them "*ignoramus et ignorabimus*." Things we do not know and we will never know.

In the light of my attempt to understand what questions may be beyond knowledge, it is interesting to compare my list with the seven riddles presented by du Bois-Reymond:

1. The ultimate nature of matter and force.
2. The origin of motion.
3. The origin of life.
4. The apparent teleological arrangements of nature.
5. The origins of simple sensations.
6. The origins of intelligent thought and language.
7. The question of free will.

Du Bois-Reymond believed that riddles 1, 2, and 5 were truly transcendent. The first two are very much still at the heart of the questions we considered in the first few Edges. The teleological arrangement of nature refers to the question of why the universe seems so fine-tuned for life, something that still vexes us today. The idea of the multiverse is our best answer to this riddle. The last three riddles were the subject of the previous Edge, which probed the limits of the human mind. The riddle about the origin of life is the only one that we have made some headway on. Despite amazing scientific progress in the last hundred years, the other six problems still appear to be beyond the limits of knowledge.

Hilbert was not going to admit mathematical statements to the list of du Bois-Reymond's riddles. Thirty years later, on September 7, 1930, when Hilbert returned to his hometown of Königsberg to be made an honorary citizen, he ended his acceptance speech with the clarion call to mathematicians:

For the mathematician there is no Ignorabimus, and, in my opinion, not at all for natural science either. . . . The true reason why no one has succeeded in finding an unsolvable problem is, in my opinion, that there is no unsolvable problem. In contrast to the foolish Ig-norabimus, our credo avers: Wir müssen wissen. Wir werden wissen. [We must know. We shall know.]

When he said this, Hilbert was unaware of the startling announcement that had been made at a conference in the same town of Königsberg the day before the ceremony. It turns out that mathematics does contain *ignorabimus*. A twenty-five-year-old Austrian logician by the name of Kurt Gödel had proved that it was impossible to prove whether mathematics was free of contradictions. And he went even further. Within any axiomatic framework for mathematics, there will be true statements about numbers that you can't prove are true within that framework. Hilbert's directive—"*Wir müssen wissen. Wir werden wissen.*"—would eventually find its rightful place: on his gravestone. Mathematics would have to face up to the revelation that it too had unsolvable riddles.

THE NEXT SENTENCE IS FALSE

The title of this section is true.

It was self-referential statements like the two above and the one I found in my cracker last Christmas that Gödel tapped into for his devastating proof that mathematics had its limitations.

Although sentences may throw up paradoxes, we expect statements about numbers to be either true or false. Gödel was intrigued by the question of whether you could exploit the idea of self-reference in mathematical statements. He wanted to show that within any axiomatic system there would always be true statements about numbers that couldn't be proved from these axioms. It's worth pointing out that you can try to set up different sets of axioms to try to capture how numbers work. Hilbert's hope was that mathematicians could construct a system from which we could prove all mathematical truths. Gödel found a way to scupper this hope.

He began by producing a code that assigned to every meaningful statement about numbers its own code number. The advantage of this coding is that it would allow mathematics to talk about itself.

According to Gödel's code, any axiom will have its own code number. For example, the axiom "if $A = C$ and $B = C$, then $A = B$" has its own code number. But statements that can be deduced from these axioms like "there are infinitely many primes" will also have their own code numbers. Even false statements like "17 is an even number" will have a code number.

These code numbers allowed Gödel to talk using the language of number theory about whether a particular statement is provable within the system. The basic idea is to set up the code in such a way that a statement is provable if its code number is divisible by the code numbers of the axioms. It's much more complicated than that, but this is a helpful simplification.

To express the fact "The statement 'There exist infinitely many primes' can be proved from the axioms of number theory" translates into "The code number of the statement 'There exist infinitely many primes' is divisible by the code numbers of the axioms of number theory"—a purely mathematical statement about numbers that is either true or false.

Hold your hats as I take you on the logical twists and turns of Gödel's proof. He decided to consider the statement S: "This statement is not provable." Statement S has a code number. Thanks to Gödel's coding, S is simply a statement about numbers. Either the code number of S is divisible by the code numbers of the axioms or it isn't. It must either be true or false. It can't be both, or it would contradict our assumption that the system doesn't lead to contradictions.

Suppose there is a proof of S from the axioms of number theory. This implies that the code number of S is divisible by the code number of the axioms. But a provable statement is true. If we now analyze the meaning of S, this gives us the statement that the code number of S is not divisible by the axioms. A contradiction. But we are assuming that mathematics doesn't have contradictions. Unlike my Christmas cracker paradox, there has to be a way out of this logical conundrum.

The way to resolve this is to realize that our original assumption is false—we can't prove S is true from the axioms of number theory. But that is exactly what S states. So S is true. We have managed to prove that Gödel's statement S is a true statement that can't be proved true.

One of the striking consequences of Gödel's proof is that there is no way of rescuing mathematics by making one of these unprovable statements into an axiom. You may think that if the statement S is true but unprovable, why not just make it an axiom and then maybe all true statements are provable? Gödel's proof shows why, however many axioms you add to the system, there will always be true statements left over that can't be proved.

Don't worry if your head is spinning a bit from this logical dance. Despite having studied the theorem many times, I'm always left a bit dizzy by the end of the proof. But the implications are amazing. Gödel proved mathematically that mathematics has its limitations. The intriguing thing is that the statement S is not unknowable. We've actually proved that it is true. What Gödel showed is that it can't be proved true within that axiomatic framework.

Gödel used this devastating revelation, called Gödel's first incompleteness theorem, to kill off Hilbert's hope of a proof that mathematics was free of contradictions. If you could prove mathematically that there were no contradictions, this could be used to give a proof that the statement "This statement is not provable" is true. But that is a contradiction because it says it can't be proved. So any proof that mathematics is free of contradictions would inevitably lead to a contradiction. We are back to our self-referential statements again. The only way out of this was to accept that we cannot prove mathematically that mathematics is free of contradictions. This is known as Gödel's second incompleteness theorem. Much to Hilbert's horror, it revealed *ignorabimus* at the heart of mathematics.

Mathematicians are very attached to their belief that mathematics is free of contradictions. If there were contradictions, how could we have come this far without the edifice collapsing? We call a theory consistent if it is free of contradictions. The French mathematician André Weil summed up the devastating implications of Gödel's work: "God exists since mathematics is consistent, and the Devil exists since we cannot prove it."

Did Gödel's revelations mean that mathematics was as open to falsification as any other scientific theory? We might have hit on the right model, but just as with a model for the universe or fundamental particles, it might suddenly fall to pieces under the weight of new evidence.

For some philosophers there was something attractive about the fact that, although we cannot prove Gödel's statement S is true within the axiomatic system for number theory, we have at least proved it is true by working outside the system. It seemed to imply that the human brain was more than a mechanized computing machine for analyzing the world mathematically. In a paper called "Minds, Machines and Gödel," presented to the Oxford Philosophical Society in 1959, philosopher John Lucas put forward the argument that if we modeled the mind as a machine following axioms and the deductive rules of arithmetic, then, as it churned away constructing proofs, at some point it would hit the sentence "This statement is not provable" and would spend the rest of time trying to prove or refute it, while we as humans could see that it was undecidable by understanding the content of its meaning. "So the machine will still not be an adequate model of the mind . . . which can always go one better than any formal, ossified, dead system can."

It is an attractive argument. Who wouldn't want to believe that we humans are more than just computational devices, more than an app installed on some biological hardware? In his recent research on consciousness, Roger Penrose used Lucas's argument as a platform for his belief that we need a new physics to understand what makes the mind conscious. But although it is true that we acknowledge the truth of the statement "This statement is not provable" by working outside the system, we are assuming that the system we were working within to prove the truth of Gödel's sentence is itself free of contradictions. And this takes us to Gödel's second incompleteness theorem: we can't prove it.

Gödel's statement—"This statement is not provable"—may seem a little esoteric from a mathematical point of view. Surely the really interesting statements about numbers—the Riemann hypothesis, Goldbach's conjecture, the PORC conjecture—will not be unprovable? For many years the hope remained for a number of mathematicians that only tortuous Gödelian sentences would transcend proof. But then in 1977, Jeff Paris and Leo Harrington came up with a bona fide mathematical statement about numbers that they could show was true but unprovable within the classical axiomatic setup for number theory.

If you've made it this far in this chapter, you are now just about ready to get one of the other jokes from my Christmas crackers. The only other thing you need to know to understand it is that the linguist and philosopher Noam Chomsky made a distinction between linguistic competence (the linguistic knowledge possessed by a culture) and linguistic performance (the way the language is used in communication). The joke goes like this:

> Heisenberg, Gödel, and Chomsky walk into a bar. Heisenberg looks around the bar and says, "Because there are three of us and because this is a bar, it must be a joke. But the question remains, is it funny or not?"
>
> Gödel thinks for a moment and says, "Well, because we're in-

side the joke, we can't tell whether it is funny. We'd have to be outside looking at it."

Chomsky then looks at both of them and says, "Of course it's funny. You're just telling it wrong."

14

> The eternal silence of these infinite spaces frightens me.
> —Blaise Pascal, *Pensées*

WE DISCOVERED in the Fourth Edge that there are limits to how far we can see and explore. But I spend my life exploring a mental universe of mathematical truths. Here I don't need telescopes, spaceships, or microscopes. I have different tools for probing the limits of this world. For a start, the question of infinity can be settled with the finite equipment inside my head. Mathematics allows us to transcend barriers that prevent us from exploring the edges of the physical universe. But are there limits to how far my finite neurological tools can navigate the truths in this infinite universe of numbers? Before the nineteenth century, "infinity" was synonymous with "beyond knowledge." And yet ever since the ancient Greeks concocted the dark art of mathematics, we humans have been using our finite minds to navigate the infinite.

I can't resist sharing another one of the mathematical jokes that burst out of our Christmas crackers:

> Teacher: What's the biggest number?
> Student: Seventy-three million and twelve.
> Teacher: What about seventy-three million and thirteen?
> Student: Well, I was close!

The ancient Greeks understood that numbers do not have an end, but this was different from knowing that infinity genuinely existed. Aristotle made the distinction between potential infinity and actual infinity. You can keep adding 1 to a number, but that is just a way to make a finite thing bigger and bigger. To Aristotle, that didn't actually make infinity. Even if the Greeks couldn't create an actual infinity, it is striking how nimbly they were able to use logic to explore a potential infinity.

One of the first great proofs in Euclid's *Elements* is the explanation that among the universe of numbers there is an infinite number of indivisible numbers, which we call prime numbers. Even today, this proof gives me a frisson of excitement—the idea that something that outwardly looks infinite and intractable can nevertheless be understood.

You may ask, since you've admitted to the possibility that numbers are unlimited, why should proving that there are infinitely many primes be so extraordinary? After all, knowing there are infinitely many even numbers is not so surprising once you have admitted that numbers go on forever. And yet it is the fact that we don't genuinely understand the nature of primes that makes the proof so surprising. It shows that the set of primes is infinite without being able to say what they are. The question of whether the physical universe is infinite might require a similar approach: a logical argument implying that it must go on forever without our physically being able to see it.

For millennia, infinity has caused problems. Most regarded it as an expression of something beyond our comprehension. Thomas Aquinas, the thirteenth-century Christian theologian and philosopher, wrote:

> The existence of an actual infinite multitude is impossible. For any set of things one considers must be a specific set. And sets of things are specified by the number of things in them. Now no number is infinite, for number results from counting through a set of units. So no set of things can actually be inherently unlimited, nor can it happen to be unlimited.

Discussions of infinity were never too far removed from questions of theology. In the fifth century, St. Augustine wrote in *City of God* that infinity was reserved for the mind of God. He was dismissive of those who believed infinity was beyond even the mind of God:

Such as say that things infinite are past God's knowledge may just as well leap headlong into this pit of impiety, and say that God knows not all numbers. . . . What madman would say so? . . . What are we mean wretches that dare presume to limit his knowledge?

The medieval philosopher Oresme, who contemplated the idea that there might be infinite space beyond the celestial sphere that encloses our universe, was also adept at manipulating mathematical infinities. He was the first to prove the startling fact that if you keep adding up the fractions $1 + \frac{1}{2} + \frac{1}{3} + \frac{1}{4} + \ldots$ the answer can be as big as you want. He was also one of the first to consider the idea that you might be able to compare the size of different infinities. After all, if you compare the infinity of all whole numbers and the infinity of even numbers then you could match every whole number with its double. Yet Oresme concluded that comparing infinities was a dangerous move.

For centuries many felt that arguments of this sort proved that infinity could not really exist. The fourteenth-century English cleric and mathematician Thomas Bradwardine used a similar logic to prove that the world could not be eternal. If the world were eternal, he argued, the number of female souls and the number of all souls would be infinite. If they were infinite, you could match them up. But that would leave no room for the male souls. Therefore, an infinity of souls led to a contradiction.

Galileo confronted a similar problem when contemplating the number of square numbers. On the one hand, there are clearly more numbers that are not square numbers than numbers that can be written as a square. The square numbers 1, 4, 9, 16, 25, . . . get sparser and sparser, with more non-squares appearing between each successive square. And yet wasn't every number the square root of some square number? So from this perspective every number could be paired with a square number, leading to the conclusion that there were the same number of squares as of numbers.

Like Oresme before him, Galileo was flummoxed by this. As he wrote in the *Two New Sciences*:

> **The difficulties in the study of the infinite arise because we attempt, with our finite minds, to discuss the infinite, assigning to it those properties which we give to the finite and limited; but this . . . is wrong, for we cannot speak of infinite quantities as being the one greater or less than or equal to another.**

Shortly after Galileo's death, the now familiar symbol that represents infinity was introduced. The symbol ∞ was first used by the English mathematician John Wallis in 1655. He chose it to capture the idea that one could traverse the curve infinitely often.

For another two centuries, mathematicians were happy with the idea of a potential infinity but not with the idea that infinity genuinely existed, which just seemed to cause too many problems. As the nineteenth-century mathematician Carl Friedrich Gauss wrote in a letter to his colleague Heinrich Christian Schumacher, "I protest against the use of infinite magnitude as something completed, which is never permissible in mathematics. Infinity is merely a way of speaking."

Then, at the end of the nineteenth century, an intellectual shift occurred. Thanks to one man's finite brain, the infinite suddenly seemed to be within reach. For the nineteenth-century German mathematician Georg Cantor, infinity was not simply a way of speaking. It was a tangible mathematical object: "The fear of infinity is a form of myopia that destroys the possibility of seeing the actual infinite, even though it in its highest form has created and sustains us, and in its secondary transfinite forms occurs all around us and even inhabits our minds."

At the end of the nineteenth century many practicing scientists questioned the basic facts of religion. But religious Georg Cantor embraced science and religion, and he wrote with clarity about the influence of his faith on his mathematical ideas. His belief in God was the foundational hypothesis from which he deduced that the infinite must exist: "One proof is based on the notion of God. First, from the highest perfection of God, we infer the possibility of the creation of the transfinite, then, from his all-grace and splendor, we infer the necessity that the creation of the transfinite in fact has happened."

Cantor's proposal for navigating the infinite goes back to some of the ideas contemplated by Oresme. Cantor suggested that we think of mathematicians as a tribe with names for the numbers 1, 2, and 3, but beyond that every number was just called "lots," the tribe's version of infinity. If two such tribes meet, they are still able to compare sizes and judge whose tribe is bigger. Members of the first pair up with members of the second, and whichever tribe has people unpaired has the biggest "lots." If the tribes match up exactly, then their "lots" are equal.

This is a good description of the mathematics of the animal kingdom: Animals probably don't have names for numbers. But they can still judge whose group is the biggest. Developing a sense of size is key to their survival. If one group encounters another, they need to assess quickly which one is bigger. Bigger they fight, smaller they flee. They don't need names for the numbers to make that comparison.

Using the idea of pairing objects, Cantor was able to propose a way of determining whether two infinite sets were numerically identical. You might be tempted to think, for example, that there are half as many even numbers as there are all numbers. But Cantor showed, just as Oresme had suggested, that one can line up both sets of numbers so that each number has its pair—1 is paired with 2, 2 with 4, 3 with 6, and n with $2n$. So these two sets are actually the same size. The tribe that arrives with even numbers on their backs will be able to hold its own against a tribe made up of all numbers.

Oresme and Galileo offered a similar argument, but both were disturbed by the fact that even numbers or square numbers are a subset of all numbers and should then in some sense be the smaller set. Cantor believed that you only had to find one way to pair up sets to deem them equal in size. For finite sets, if you tried pairing them up and couldn't do it, no reordering or rearranging would create a way of pairing everything up perfectly. But in the case of the infinite, Cantor saw that rearranging the order of a set might help you find a new way to pair them so that nothing was left over.

What about fractions? How big is that infinity? Cantor came up with an ingenious way to compare all whole numbers with all fractions and prove that both sets are identical in size. At first sight, this looks impossible: between each pair of whole numbers there are infinitely many fractions. But there is a way to match all whole numbers perfectly with all fractions so that no fractions are left unmatched.

It starts with a table containing all the fractions. The table has infinitely many rows and columns. The nth column consists of a list of all the fractions $1/n, 2/n, 3/n, \ldots$

How did Cantor pair up the whole numbers with the fractions in this table? The trick is, first of all, to wend a snake diagonally through the fractions in the table, as illustrated on the next page. Then, in order to pair the whole numbers with all the fractions, what we do is work our way along the path, pairing 1 with ⅟₁, 2 with ⅔₁, 3 with ½, 4 with ⅓. The number 9, for example, gets paired with ⅔, the ninth fraction that we meet as the snake slithers through the table of fractions. Since the snake covers the whole table, every fraction will be paired with some whole number.

$$
\begin{array}{ccccc}
\frac{1}{1} & \frac{1}{2} & \frac{1}{3} & \frac{1}{4} & \frac{1}{5} \cdots \\
\frac{2}{1} & \frac{2}{2} & \frac{2}{3} & \frac{2}{4} & \frac{2}{5} \cdots \\
\frac{3}{1} & \frac{3}{2} & \frac{3}{3} & \frac{3}{4} & \frac{3}{5} \cdots \\
\frac{4}{1} & \frac{4}{2} & \frac{4}{3} & \frac{4}{4} & \frac{4}{5} \cdots \\
\frac{5}{1} & \frac{5}{2} & \frac{5}{3} & \frac{5}{4} & \frac{5}{5} \cdots \\
\vdots & \vdots & \vdots & \vdots & \vdots
\end{array}
$$

It's a beautiful and surprising argument. If I were cast away on a desert island and allowed to take only eight theorems with me, Cantor's snake would be one of them. How remarkable to find a way to match up all fractions with whole numbers and show they have the same order of magnitude!

COUNTING INFINITIES

It is beginning to look like all infinities are identical in size. Perhaps once a tribe has infinitely many members it will never be beaten by another tribe. Now enter the new big-cheese tribe whose members are labeled with all the possible decimal expansions there are of numbers. Will the tribe whose members are labeled with the whole numbers 1, 2, 3, . . . be able to pair themselves up with this new tribe? I might start by matching tribe member number 1 with tribe member π = 3.1415926 . . . , then tribe member number 2 with e = 2.7182818 . . . But how can I find a way to exhaust all the members of the infinite decimal tribe? Is there a clever way to arrange the infinite decimal numbers so that I can snake whole numbers through them as Cantor did with fractions?

Cantor produced a clever argument to show why he could be certain that, however hard I tried to match up the whole-number tribe, he could guarantee that an infinite decimal tribe member would always be unaccounted for. The infinity of all infinite decimal expansions of numbers is a genuinely bigger sort of infinity than the infinity of whole numbers. It's a beautifully simple if subtle argument and I think it would be another one of the mathematical theorems I'd take to my desert island.

How could Cantor be sure that he could always guarantee an infinite decimal tribe member would be unaccounted for? Let's take one of my attempts to match the whole-number tribe with the infinite decimal numbered tribe.

$$1 \leftrightarrow 3.1415926\ldots$$
$$2 \leftrightarrow 2.7182818\ldots$$
$$3 \leftrightarrow 1.4142135\ldots$$
$$4 \leftrightarrow 1.6180339\ldots$$
$$5 \leftrightarrow 0.3331779\ldots$$
$$\ldots$$

What Cantor does is to build a number with an infinite decimal expansion that he can guarantee is not on my list and has not been paired up with one of the whole numbers. He does this by choosing each decimal place of this new number one step at a time, working his way down the list. Each decimal place is a number between 0 and 9. In the first decimal place, Cantor chooses a number that is different from the first decimal place of the number paired with number 1. In the second decimal place, he chooses a number different from the second decimal place of the number paired with number 2. Cantor carries on building this new infinite decimal number, choosing a different decimal place each time as he works his way through the list.

$1 \leftrightarrow 3.1415926\ldots$

$2 \leftrightarrow 2.7182818\ldots$

$3 \leftrightarrow 1.4142135\ldots$

$4 \leftrightarrow 1.6180339\ldots$

$5 \leftrightarrow 0.3331779\ldots$

$\ldots$

There are lots of different ways he could cook up this new infinite decimal number, but one example might start $0.22518\ldots$, which we can already see is not paired with the first five whole numbers because the infinite decimal is different from the first five infinite decimals in this list. But how can we guarantee that this number isn't somewhere in the list we've made? Perhaps it's paired with, say, number 101. Given the way this new number has been constructed, Cantor could simply say, "Check the 101st decimal place; it's different from the 101st decimal of this new number." The number has been cooked up so that it can't be anywhere on this list. That means there are decimal numbers missing from the list.

There are a few technical points to watch out for, but the essence of the argument suffices to show that there are more numbers with infinite decimal expansions than there are whole numbers. The argument works for any attempt I might make to pair up whole numbers with infinite decimal numbers—there are always infinite decimals left over.

Cantor himself was genuinely surprised by his discovery. "I see it, but I do not believe it," he said.

He eventually proved that there are infinitely many types of infinity. He showed we could replace "lots" with meaningful names for all these different infinities and introduced new symbols for these new infinite numbers, using letters of the Hebrew alphabet. The smallest infinity, $\aleph_0$, was called aleph-zero, after the first letter of the Hebrew alphabet. Cantor was probably aware of its mystical significance in

Jewish Kabbalah, where it represents the infinity of God, but the choice also hints at the idea of a new beginning. For me, this is one of the most exciting moments in mathematical history. It is like the moment when we counted for the first time. But instead of counting 1, 2, 3, we were counting infinities.

The great German mathematician David Hilbert recognized that Cantor had created a genuinely new mathematics. Hilbert declared Cantor's ideas about infinities to be "the most astonishing product of mathematical thought, one of the most beautiful realizations of human activity in the domain of the purely intelligible . . . no one shall expel us from the paradise which Cantor has created for us." I would agree.

Cantor believed that he was tapping into the mind of God, and that he was just a mouthpiece for God. It may be that his faith gave him the courage to believe that infinity also existed. But it was Cantor's mathematical brilliance that wrested the infinite from the unknown into the known. The Christian Church, far from being disturbed by Cantor's attempts to read the infinite mind of God, took an active interest in his emerging ideas. Cantor entered into a lengthy correspondence with cardinals at the Vatican over the nature of God and infinity.

But not everyone was so enamored with his ideas. One of the most influential mathematicians in Germany, Leopold Kronecker, thought Cantor's mathematics an aberration. "I don't know what predominates in Cantor's theory—philosophy or theology," he wrote, "but I am sure that there is no mathematics there." Kronecker once famously said, "God created the integers. All the rest is the work of man." But what Cantor had created was so revolutionary that Kronecker regarded it as a mathematical carbuncle. Kronecker's opposition ensured that Cantor never claimed a job at one of the big German universities, including Berlin, where Kronecker was based. Instead, he spent his days in the backwater of the University of Halle. Cantor tried to fight back, writing directly to the minister of education, but to no avail.

Even publishing his ideas proved difficult. Another influential mathematician of the day, Gösta Mittag-Leffler, eventually rejected Cantor's work, declaring it a hundred years ahead of its time. Cantor was severely thrown by this rejection from a mathematician he greatly respected. The continual battle with the establishment, his struggles with the mysteries of the infinite, and the deaths of his mother, brother, and then youngest child took their toll. Cantor suffered from bouts of manic depression, and the controversy only exacerbated his condition. He was admitted to the Nervenklinik, a mental hospital in Halle, in 1884 and spent much of the last few decades of his life there. Disillusioned with mathematics, he turned to questions of religion. He also spent a lot of time trying to prove that Francis Bacon was the real author of Shakespeare's plays.

Mittag-Leffler's premonition was right in some respects. A century later, Cantor's ideas are considered some of the most beautiful and extraordinary proofs of the last three hundred years. Cantor has allowed mathematicians to touch infinity, to play with it, compute with it, to finally recognize that infinity is a number. Not just one number, but infinitely many numbers.

But for Cantor, the infinite was not just an abstraction:

I am so in favor of the actual infinite that instead of admitting that Nature abhors it, as is commonly said, I hold that Nature makes frequent use of it everywhere, in order to show more effectively the perfections of its Author. Thus I believe that there is no part of matter which is not—I do not say divisible—but actually divisible; and consequently the least particle ought to be considered as a world full of an infinity of different creatures.

YES AND NO

Cantor's discovery of these layers of infinity led to a very real example of a problem that could not be resolved within the current axioms of mathematics—a statement without proof, undecidable, beyond what we can know. It was a question that went to the heart of what a number is and revealed just how subtle numbers are.

Cantor wanted to know whether there are sets of numbers that are bigger in size than whole numbers but small enough that they can't be paired with all infinite decimal expansions. In other words, can there be a tribe that beats the whole-number tribe but is outdone by the tribe with all infinite decimal numbers? The name given to the infinity of all infinite decimal numbers is the continuum. The continuum hypothesis posited that there was no infinity smaller than the continuum but larger than the infinity of all whole numbers.

Hilbert was so struck by the challenge of proving the continuum hypothesis that he put it at the top of his list of twenty-three problems for the mathematicians of the twentieth century.

It is a question Cantor wrestled with his whole life. One day he was convinced he'd proved there was no infinity between these two. But then he found a mistake. The next day he thought he'd proved the opposite. As Cantor always believed, "In mathematics the art of asking questions is more valuable than solving problems."

And so it turned out to be. The reason Cantor was having so much trouble was that both answers are correct.

The answer to this problem, which finally arrived in the 1960s, rocked the mathematical community to its foundations. Building on work done by Gödel, Paul Cohen, a logician at Stanford, demonstrated that you couldn't prove from the axioms we currently use for mathematics whether there was a set of numbers whose size was strictly between the number of whole numbers and all infinite decimal numbers. In fact, he produced two different models of numbers that satisfied the axioms we use for mathematics; in one model the answer to Cantor's question was yes, and in the other model the answer was no.

I'm not sure whether Cantor would have liked this conclusion. He did once declare, "The essence of mathematics lies entirely in its freedom."

Some regard this as similar to the moment when mathematicians discovered that there are many sorts of geometries in addition to Euclid's. Euclid's geometry satisfies the parallel postulate, while the new spherical and hyperbolic geometries don't. Now we understood that there are different models for our numbers, some of which have these intermediate infinities, others of which don't.

It was deeply unsettling for many mathematicians. We thought we knew our numbers. Numbers like the square root of 2 and π might be irrational, with infinite decimal expansions, but we felt that we could see these numbers marked on a ruler. So in the case of the numbers we think we know, we feel there should be an answer to Cantor's question. Is there a subset of numbers on this ruler that is strictly bigger than the whole numbers, yet strictly smaller than the set of all infinite decimal numbers? Most mathematicians believe that the answer must be yes or no, not yes and no.

Cohen's colleague Julia Robinson wrote to him, "For heaven's sake there is only one true number theory! It's my religion." Intriguingly, she crossed out the last sentence before sending the letter. Cantor himself would probably in the end have embraced this new uncertainty, as his religion encouraged the acceptance of things that transcended human knowledge.

How many other unresolved problems on the mathematical books will turn out to be unprovable? To tackle some of these great unsolved problems we may need to add new axioms. Gödel thought this could account for the difficulty of proving the Riemann hypothesis, the greatest unsolved problem of mathematics. He questioned whether our current axioms were sufficient to tackle many of the problems of number theory:

> One is faced with an infinite series of axioms which can be extended further and further, without any end being visible. . . . It is true that in the mathematics of today the higher levels of this hierarchy are practically never used . . . it is not altogether unlikely that this character of present-day mathematics may have something to do with its inability to prove certain fundamental theorems, such as, for example, Riemann's hypothesis.

PROVING WHAT IS TRUE

Gödel's incompleteness theorem is a fascinating microcosm of the challenge of trying to prove what is true. Within any consistent axiomatic system for number theory there will be truths that cannot be proved true. The intriguing thing is that by working outside the system we can prove that a particular statement is true but unprovable within the system. You might say: Then why not work in that larger framework? But Gödel guarantees that this larger system will have its own unprovable true statements, requiring another move outside the system. It's a very familiar infinite regress.

There are resonances with many of the problems I have been wrestling with. Perhaps it is impossible to understand the universe when we are part of the system. If the universe is described by a quantum wave function, does something need to be outside the system to observe it? Chaos theory implies that we can't understand part of the system as an isolated problem, because an electron on the other side of the universe may have an influence that will send a chaotic system in a completely different direction. We must stand outside in order to consider the whole system. The same problem applies to the question of understanding consciousness. We are stuck inside our own heads, our own systems, unable to access anyone else's consciousness. This is also why some say we will always be limited in our ability to tackle questions like the existence of a God that transcends the universe we are stuck in.

The other important aspect of Gödel's work is the fact that you cannot prove that mathematics is consistent, that it contains no contradictions. This begs the question of whether the methods we employ to gain knowledge about the universe are effective. Any attempt to explain, for example, why induction is the right strategy for studying physical phenomena is going to rely on applying induction. The whole thing becomes very circular.

Cohen's discovery led to the exciting revelation that there are statements about numbers that you can either assume are true or assume are false and both answers derive from legitimate models. Can we do the same once we encounter a genuinely unknowable question about the universe? If we identify a question whose answer cannot be known, then it makes perfect sense to work under the hypothesis that the answer is one thing or another. Your choice of working hypothesis may depend in part on the probability that the answer is one thing rather than another. But sometimes probabilities are not relevant, and your choice may be based on your personal relationship to the consequences that follow from working within that system.

In mathematics we are freed up from this need to choose. As a mathematician, I'm quite happy to move between different mathematical models that are individually self-consistent yet mutually contradictory. For example, I can do mathematics under the assumption that the continuum hypothesis is true or that it is false. If the original model is consistent, then both mathematical models are consistent. If it makes my equations work, I might use the continuum hypothesis to explore this particular mathematical universe. Can we do the same thing with other unknowables? If God is the unknowable, can we make a choice to give flesh to this unknown? But by doing that, aren't we trying to make it knowable, going against the spirit of the definition?

You have to be careful when making choices or hypotheses under which to operate. You can't just assume that something you don't know can be true or false. You have to prove first that both possibilities are consistent. For example, I don't know whether the Riemann hypothesis about prime numbers is true or false, but only one of these is consistent with our theory of numbers. If it turns out to be unprovable within the current axioms of number theory, this would actually imply that it was true. If it is false, we know that it is provably false because we will be able to find a counterexample with a finite systematic search. If it is false, you can't work within a model of numbers where you assume it is true because you'll generate contradictions. This is the amazing thing about the continuum hypothesis—either it or its negation can be incorporated into a consistent theory without creating contradictions.

Some have argued that the numbers we are trying to capture axiomatically are meant to be the numbers we measure with, that are located somewhere on a ruler. So there might be good reason to assume the continuum hypothesis as a better description of what we are trying to model. In fact, logician Hugh Woodin has recently put forward arguments for why the numbers we are trying to model are those in which the continuum hypothesis is false. He argues that if these numbers model measurements on a ruler, then there are reasons to believe this would imply that there are infinite subsets of intermediate size between the set of whole numbers and the set of infinite decimals.

This illustrates the tension between mathematics and physics. For centuries, mathematics has been happy with the mathematical multiverse: different, mutually exclusive mathematical models of geometry or number theory. But even if the physicist is happy with the idea of the multiverse, there is still the desire to identify which of these possibilities describes the universe in which we exist.

Suppose a scientist cooks up a perfectly consistent logical theory of how the universe might work, but then discovers that it doesn't match the experimental evidence of our own universe. The hypothetical theory is booted out and is no longer of interest to the world of science. If a biologist starts writing papers about a hypothetical beast that potentially could exist, like a unicorn, but doesn't actually walk this Earth, no one is going to be interested unless it casts some light on the animals that do exist. By contrast, in mathematics such new worlds and beasts are celebrated and embraced. They become part of a richer tapestry of mathematical possibilities. Science does the actual, mathematics does the possible. Science charts a single pathway through a tree of possible universes, mathematics maps every possible journey.

But what do we do about science questions that can't be resolved? In the case of physics, once we have a question that is beyond knowledge, one answer will still be a correct description of our universe, and one will not. But by its nature, if it is unanswerable, we are not going to get any new evidence about our universe that would allow us to decide which is the true description and which the false. If we do, the question wasn't unanswerable in the first place. So what happens to those who labor under the false hypothesis? Nothing! As with the continuum hypothesis, its negation will still be consistent with our current theory of the universe. You will get a different story with different theorems and results. If it led to contradictions, this would give us a way to know that it was wrong, and hence the question was not genuinely unanswerable.

Consider the question of whether the universe is infinite. If this is an unknowable, it is because the universe is infinite or because it is finite and of such a size that it will always remain beyond our event horizon and therefore beyond the realms of investigation. If it is infinite, what happens to someone who works under the hypothesis that it is finite but too big ever to prove finite? The interesting thing is that if this led to problems—it contradicted current theory or new data— we'd have a way of proving that the universe is infinite. Of course, in mathematics this is a great way to proceed. We proved that the square root of 2 cannot be captured by a ratio of two finite numbers by working under the hypothesis that it could and eventually running up against a contradiction. It is possible that an infinite universe might be the only assumption that doesn't lead to contradictions. It could be that mathematics is again our best means of exploring the far reaches of the universe.

In this Edge I've explored how mathematics has even managed to help us navigate infinity itself. It is quite striking that infinity was once regarded as something beyond knowledge and often connected to the idea of God. Descartes wrote, "God is the only thing I positively conceive as infinite." And yet Cantor's extraordinary insights at the end of the nineteenth century provide us with a way to explore and compare infinities. Infinity is no longer beyond our reach. Cantor was not unaffected by the implications of his discoveries for the question of God. Indeed, he believed he had been chosen by God to reveal these ideas about infinity to the world.

Infinity has played a key role in the exploration of the existence of God. Thomas Aquinas's proof of God, known as the cosmological argument, proposes that anything that exists needs a creator, but this would then apply to the creator itself. To avoid infinite regress, God is the solution to the problem of a first cause. But mathematics allows us to keep building new infinities by considering all the old infinities and then taking their union, so there is no need to stop this chain of creators, as Aquinas believed. Each time we get something new, and the process never ends.

Even though the mathematics is there to keep building new infinities from the old, it becomes very hard even for a mathematician to grasp the extent of our mathematical universe. Most content themselves with playing down in the lower echelons of the infinite. But we know that these are just part of an unending hierarchy. This gives those who try to define God as "that which nothing greater can be thought of" some problems. In one sense, such a thing doesn't exist because there is always a way to make something greater. But if we come back to the idea of something beyond that which humans can conceive of, then we are back to things we as humans cannot know: the limitations of our own biology.

BEYOND
THE EDGE

S O, AT THE END of my journey to the edges of knowledge, have I
found things that we can categorically say we cannot know? The
things I thought we could never know, such as whether the universe
is infinite, turn out not to be as unassailable as I thought. We can use
mathematics to employ finite means to prove the existence of the infinite. So although we may never explore or see beyond the finite bubble that encloses our bit of the universe, we may be able, using the
power of our minds alone, to discover what is beyond.

Understanding the nature of time before the Big Bang was another edge that I thought was unassailable. But chinks have opened up
in that wall, too. Recent progress has provided us with ways to theo-
rize and even perhaps detect evidence of a time before we thought it
all began. And yet the question of whether time has a beginning feels
like one that will remain on the scientific books for some time yet.

In contrast, the infinitely small at the heart of the die that I hold
in my hand feels like something that will always remain beyond complete knowledge. Every generation imagines they've hit the indivisible, only for matter to fall apart into smaller pieces. How can we ever
know that the current building blocks of the universe—quarks, elec-

trons, neutrinos—aren't as divisible as all the other particles we've hit as we've peeled the onion of reality? Quantum physics even posits a limit to how far we will be able to penetrate in our investigation of matter: beyond the Planck length, it's simply no-go.

Whether we can ever truly understand the nature of consciousness is very much in a state of flux. Will this challenge vanish because it will turn out to be an ill-posed question? Will it be answered with a strategy similar to that used by scientists to pin down the essence of life? There was no *elan vital*, just a set of biological processes that means a collection of molecules has life. Or will the problem of consciousness remain something that can never be understood because we are stuck inside our own consciousnesses and can never get inside another's?

The possibility that we cannot know because we are stuck within the system is a common theme among many of the problems I have tried to tackle. Mathematics has truths that will remain unprovable within a system. Step outside and you can know, but this creates a new system, which will have its own unprovable truths. Likewise, the idea that a quantum experiment can be repeated is an impossibility because we can never isolate the experiment from the universe it is conducted in, one that will have changed and evolved by the next time we run the experiment.

Even the mathematics of probability is something of a fantasy. What is probability? If I throw my die six hundred times, I expect to get one hundred 6s. But I just want to throw it once and know something about how it will fall. The equations of chaos theory tell us that so much of the future depends on fine-tuning the decimal places that control the input of the equations. So I can never know the present completely enough to have any real chance of knowing the future or the past.

The physical limitations of the human brain place boundaries on how much we can ever know, so there will always be things beyond knowledge. But this isn't an absolute unknowability. It is more like light coming in from the outer reaches of the universe before we realized that the expansion of the universe was accelerating. Wait long enough and the light will arrive. Wait long enough and a computer can make its way through all the provable truths of mathematics. But what if time itself runs out while we are waiting?

The limitations of language are at the heart of many of the boundaries of knowledge, and these could evolve and change. Certainly, many philosophers identify language as a problem when it comes to the question of consciousness. Understanding quantum physics is also a problem because the only language that helps us navigate its ideas is mathematics. Try to translate the mathematics into the language of everyday experience and we create the absurdities that make quantum physics so challenging. So the unknowability of position and momentum isn't really a genuine unknowable. Rather, it is a failure of translation from mathematics to natural language.

We are bound by the ways of thinking particular to our own moment in history. Auguste Comte thought we could never know the constituents of a star. How wrong he was! So I wonder if the safest bet is to say that we can never truly know for sure what we cannot know.

IS GOD AN IMAGINARY NUMBER?

How far can we get by creating solutions to seemingly unanswerable problems? For many centuries, mathematicians looked at the equation $x^2 = -1$ and believed it had no answer. But then a more imaginative approach was taken. We admitted imaginary numbers into the landscape of mathematics by defining i as the number whose square is -1. Why does this work? Because it doesn't create a contradiction in the theory. We interweave this concept with the mathematics we know and we begin to know i. And most important, it gives us access to new and exciting bits of the mathematical world. Not admitting imaginary numbers into mathematics would have limited its extent and power.

What happens if I try to be creative with some of our unknowable questions? What if, for example, I were to define God as the solution to the question "Why something rather than nothing?" This concept is meant to be nothing more than the solution to that question. It doesn't have any other properties. Even if we gain more knowledge about the answer to the question, it will just mean that we know more about this particular God.

But we have to be careful with such an approach. Just because you can write down a mathematical equation doesn't mean it has solutions. Admitting a new concept that solves $x^2 = -1$ was profitable because it provided access to new consistent mathematics. For the Platonist, the idea was sitting there all along, waiting to be articulated; for others, it was a creative act that enriched our mathematical world. But if I write down the equations for Fermat's Last Theorem and try to define new numbers to solve these equations, I am going to find myself with self-contradictory statements. This is, after all, precisely how Wiles proved that the equations couldn't be solved.

The trouble with most religions is that the God that is served has so many properties that have nothing to do with the definition. It's as if we are working backward, focusing on the strange properties conjured up over the generations without really understanding the original definition. We come across this bastardized picture early on as kids, and then when we ask the question "Why something rather than nothing?" it doesn't really work as a solution. But we've been shown the wrong thing.

Being an atheist means, for me, that I reject the classical solutions that religion seems to offer for these unknowns. But maybe I shouldn't throw everything out. There are things that will always remain unknown, so perhaps God does exist. The traditional argument against the God of the gaps is that we should strive to know God, to have a personal relationship with it. And this God, defined as the transcendental or unknown, precludes by its definition the possibility of knowing it.

The trouble with this definition of God is that it doesn't really get you much further. While defining a number whose square is –1 resulted in a rich array of consequences, defining something as the solution to "Why something rather than nothing?" doesn't give rise to anything new. You need to make up properties for this thing that don't follow from its definition. As Karen Armstrong put it, this high God is too high.

There are several responses to identifying the unknowable. One is to leave it at that; if it is unknowable, then it can't be known. But there is also the temptation to make a choice and live your life according to that choice. Perhaps the logically most consistent response is to be open to many solutions and allow all the solutions to run parallel until new ideas collapse the possibilities.

I wonder, though, whether, as I come to the end of my exploration, I have changed my mind about declaring myself an atheist. With my definition of a God as that which we cannot know, to declare myself an atheist would mean that I believe there is nothing we cannot know. I don't believe that anymore. In some sense I think I have proved that this God does exist. The challenge now is to explore what quality this God has.

My statement about being an atheist is really just a response to the rather impoverished version of God offered by most religions and cultures. I reject the existence of a supernatural intelligence that intervenes in the evolution of the universe and in our lives. This is a rejection of the God that people assign strange properties to—such as compassion, wisdom, love—that make no sense when it comes to the idea that I am exploring.

Such a position and definition will probably not satisfy either side of the divide. The militant atheists don't want to admit anything named God into the debate, while those who believe in a God will criticize the concept of God as the unknown as impotent and missing the point. So how do we engage with this God of the gaps?

Perhaps the important lesson is to maintain a schizophrenic state of mind. A multi-mind-set. On the one hand, as humans we must recognize that we cannot know it all. There are provable limits to knowledge. Such a state of humility is intellectually important, or we will live in a state of delusion and hubris. Yet the other lesson is that we cannot always know what will forever transcend our understanding. This is why it is essential for a scientist not to give in too early. To believe that we can find answers. To believe that perhaps we can know it all.

DO I OWN AN ODD OR EVEN NUMBER OF DICE?

My journey through science has thrown up a number of challenging questions. But there is also the basic epistemological question of whether we can actually know anything at all. More than two thousand years ago Socrates declared, "True knowledge exists in knowing you know nothing." An acknowledgment of your ignorance is the only true statement of knowledge.

There have been volumes of philosophy that tackle the theory of knowledge and try to pin down what we can know, to define what we mean by knowledge. Plato proposed that knowledge should be defined as "justified true belief," but Bertrand Russell and then the American philosopher Edmund Gettier in the 1960s questioned whether this truly captures its meaning.

The classic example proposed by Bertrand Russell tells the story of a woman who looks at a clock that says two o'clock. She believes therefore that it is two o'clock. She seems justified in her belief because the clock says so. And it does indeed happen to be two o'clock. Yet actually the clock stopped twelve hours ago, and it's just a coincidence that she happened to look at the clock exactly twelve hours later.

Gettier created similar scenarios to challenge "justified true belief." You are looking at a field and see what you believe is a cow. You infer that there is a cow in the field. Actually, there is a cow in the field, so the inference is true, but it can't be seen because it is in a dip in the field. You are in possession of a true statement. It was based on a justifiable belief, and the thing you were looking at certainly looked exactly like a cow. But the fact that the statement you made is true does not imply knowledge.

We can imagine a situation in which we have come up with a true statement about the universe. But the justification for the statement is completely incorrect, even if it led us to make a true statement. Surely that doesn't constitute knowledge. I have often cooked up proofs of true mathematical statements that turn out to have a logical flaw in them (which I hope to spot before I send them to a journal for publication). But my false proof can't really justify my knowledge that the mathematical statement is true.

I don't know whether the Riemann hypothesis is true or false. However, a few people have come up with what they believe are proofs of the truth of this hypothesis, and they have pages and pages of equations to back up their belief. Most of the time a fault is found. Once shown to the proposer, that justified belief vanishes. But what if that false proof convinces everyone? Suppose the fault is actually quite subtle. We cannot say we know that the Riemann hypothesis is true despite our justified true belief. Surely the justification has to be true to lead to justified true belief.

Some ancient astronomers proposed that the Earth goes around the sun, but their justification of this fact was not correct. The Indian philosopher Yajnavalkya justified this belief in a heliocentric solar system in the ninth century BC by proposing, "The sun strings these worlds—the Earth, the planets, the atmosphere—to himself on a thread." Can we say that he knew that the Earth goes around the sun?

I think that I side with my colleague in New College, Timothy Williamson, who asserts in his book *Knowledge and Its Limits* that knowledge should be regarded as something fundamental, not something that can be defined in terms of other things. It seems like we all seem to know what "to know" means. It is one of only a hundred or so phrases that have a comparable translation in every language on Earth, which is not the case for as basic a word as *eat*.

It was also from Williamson that I learned about a fantastic piece of logical trickery called the paradox of unknowability, which proves that, unless you know it all, there will always be truths that are by their nature unknowable. This paradox is attributed to American logician Frederic Fitch, who published it in a paper in 1963. Fitch admitted that the original source of the argument was actually a comment made in 1945 by an anonymous referee about a paper Fitch had submitted for publication but had never made it into print. For many years, the referee responsible for this logical gem remained a mystery. But subsequent detective work has tracked down the handwritten copy of the report, and an analysis of the handwriting reveals the author to be the famous American logician Alonzo Church, who made major contributions to the understanding of Gödel's incompleteness theorem.

Church's argument has a whiff of the self-referential strategy that Gödel employed, but this time there is no mathematics involved, just pure logic. And while Gödel proves that there are mathematical truths that can never be proved within a particular consistent axiomatic system of mathematics, Church goes one further, promising a truth that will never be known by any means.

Suppose there are true statements that I do not know are true. In fact, there are lots of such statements. For example, my house is full of dice, not just the casino die that I have on my table. There are dice in our *Monopoly* set, our box of ludo, dice that have gone missing down the side of our sofa, dice that are buried in the chaos of my kids' rooms. I do not know whether my house contains an odd or even number of dice. This by itself is, of course, not an unknowable statement, because I could make a systematic search of my house to determine the answer. But it is certainly something that, at this point in time, I do not know the answer to.

Now hold on tight—this is the bit that leaves my brain in a spin every time I read it. Let p be the true statement between the two options: "there are an even number of dice in my house" and "there are an odd number of dice in my house." I don't know which one is true, but one of them must be. The existence of an unknowable truth is squeezed out of the existence of this unknown truth. The following statement is an unknowable truth: "p is true but unknown." It is certainly true. Why is this unknowable? Because to know this means I know that p is true and unknown, but that's a contradiction because p can't be unknown and known simultaneously. So the statement "p is true and unknown" is itself an unknowable statement. It's not that p itself is unknowable. As I said, I can go and find all the dice in my house and know whether it is an odd or even number. It is the meta-statement "p is true and unknown" that is unknowable. The proof works, provided something exists that is true but unknown. The only way out of this is if I already know it all. The only way that all truths are knowable is if all truths are known.

Although it has become known as a paradox, there is, as Williamson points out, no paradox involved. It is simply a proof that there are unknowable truths. After our long journey to the outer limits of science, it turns out that a clever logical riff produces the answer I was looking for.

CAN WE KNOW ANYTHING?

Many philosophers of knowledge question how much we can ever really know about anything. The eighteenth-century Scottish philosopher David Hume identified one of the fundamental problems we've had with many of the questions I've been tackling—that of being stuck inside the system. If we are going to apply scientific methods to establish that we actually know something, we get into a loop because we are using scientific and logical arguments to prove that these methods are sound. It is impossible to assume an outside position. Wittgenstein summed this up colorfully: "You cannot shit higher than your arse."

What about mathematics? Surely there we have certain knowledge. Doesn't proof give us complete certainty that, for example, there are infinitely many prime numbers? Yet even mathematical proofs have to be processed by the human brain to check whether they are correct. What if we are all convinced of the truth of an argument that nonetheless has a subtle hole in it? Of course, one of the things that we take advantage of is the fact that any fatal hole should eventually reveal itself. But doesn't this imply that mathematics, like science, is subject to an evolutionary process? The mathematical philosopher Imre Lakatos believed so. He developed a philosophy of mathematics that was modeled on Popper's view of science as something that could only be falsified, not proved true. For Lakatos, it can never be known whether a proof might hide a subtle flaw.

His book *Proofs and Refutations* presents a fascinating discussion among students exploring the proof of Euler's theorem on the relationship between the number of vertices, edges, and faces of a three-dimensional polyhedron. It mirrors the history of the evolution of this theorem: $E = V + F - 2$. At first, the students think they've got a proof. Then a student proposes a shape with a hole in the middle. The formula doesn't work on this shape. Nor does the proof. One interpretation is that the proof works for the shapes that it was intended to work on. But now a new proof and theorem are introduced pertaining to a new formula that in addition to vertices, edges, and faces, includes the number of holes in the shape. The story reveals a much more evolutionary approach to mathematical knowledge than many mathematicians will dare admit to, more akin to the process of scientific investigation. So how effective is either in discovering the truth?

One of the reasons for believing that science is producing true knowledge is its success rate. Science is so successful in its description and prediction of the way things appear to be that we feel like it must be getting close to a reality that most of us believe does exist. The fact that science works so well at making predictions and explaining phenomena is perhaps the best measure that we are close to the truth. If the map that you are using consistently gets you to your destination, that's a good sign that the map is an accurate representation of reality.

Science has mapped the universe pretty well. Space probes land on distant planets thanks to our discoveries about the nature of gravity. Gene therapies help to tackle previously untreatable conditions thanks to our discoveries about the biology of the cell. We find our way with a GPS by exploiting our discoveries about time and space. When the scientific map doesn't work, we are ready to redraw its contours to find a description that successfully helps us navigate our environment. It is survival of the fittest theory: continued success at making predictions and controlling our environment means the theory survives. Science may not really represent reality, but nothing comes close as an alternative.

Ever since Kant, we have had to wrestle with the unknowability of "things in themselves." The limits of human perception, highlighted by how easily our senses can be tricked, raise questions about how much our brain can really know about reality. Isn't everything being viewed through the spectacles we wear to look at our universe?

One of the key problems with our attempt to know the world is that we use our senses to gain knowledge of the world around us, and analytic argument to extend that knowledge. We come up with stories that match all the information that we gather with our senses. The inventions of the telescope, the microscope, and the fMRI scanner have extended how much we can perceive with our senses.

Yet what if there are things in the universe that our senses can't detect? We have more senses than many people realize; in addition to sight, hearing, taste, touch, and smell, we also have a sense called proprioception that gives us an awareness of how our body is located in space. There are also senses that give us information about the inner state of our body. The fluid in the inner ear tells us how our body's position is changing in relation to gravity. But are there physical phenomena that we miss because we don't have the sensory tools to interact with them?

Consider an organism without an eye or neurons to detect light. If it has no way of accessing electromagnetic waves, how could it ever come up with a theory of electromagnetism? We have done very well to combine our sense of sight, which can see part of the electromagnetic spectrum, with mathematical analysis to deduce other parts of the spectrum. And then we developed tools that can detect these waves and convert them into things that we can interpret. But could we have got going without being able to access, via our sense of sight, some bit of the spectrum?

It is possible that the limitations of our senses also limit the mathematics we can know. Despite mathematics being all in the mind, there is a school of thought that suggests that because our intelligence is ultimately an embodied intelligence, the knowledge we can obtain about mathematics is restricted to that which can be embodied. It is certainly true that if you look at the mathematics we do know, it often has its origins in descriptions of the physical world. Take imaginary numbers—you may question how they are embodied. And yet they emerge from the act of measuring lengths in geometric shapes. The act of understanding the diagonal across the face of my cube-shaped die led the Babylonians to consider the square root of 2. And from here we begin the journey that leads to the idea of the square root of –1.

There are proponents of artificial intelligence who assert that if we are to create an intelligence that matches ours, it must be physically embodied. In other words, a brain that lives exclusively in computer hard drives cannot generate intelligence like ours without physically interacting with the world through a body. It is a challenging hypothesis. Could there really be parts of the mathematical world that are off limits to me because they don't originate in physically embodied concepts?

And yet there is the deep philosophical issue of the extent to which our senses allow us to know anything for sure. We've already seen how our senses can be fooled into believing things that turn out to be tricks of the mind. How, for example, can we be sure that the universe as we apprehend it isn't a simulation? As we saw in the Sixth Edge, we can make someone believe they are in another person's body. So how can we be sure that we aren't just brains in a jar being fed artificial sensory information by a computer and that the whole world around us isn't just a trick?

My response to this attempt to undermine everything we know is to counter that this book has tried to explore how we can know anything about that simulation. Kant believed that the way things really are will always remain hidden from our view. All we can ever know is the appearance of things. I think most scientists spend some time reading about this debate concerning ontology and epistemology, and listening to philosophers who question whether science is really telling us how it is. And then they get back to the science, telling themselves that if we can never know what reality is really like, then let us at least try to say what the reality we apprehend through our senses is like. After all, that is the one that affects us.

So perhaps the best we can hope for is that science gives us verisimilitudinous knowledge of the universe; that is, it gives us a narrative that appears to describe reality. We believe that a theory that makes our experience of the world intelligible is one that is close to the true nature of the world, even if philosophers tell us we'll never know. As Niels Bohr said, "It is wrong to think that the task of physics is to find out how nature is. Physics concerns what we can say about nature."

But what of the things that we cannot know? If something is beyond scientific investigation, if it is unknowable, perhaps some other discipline will have a better grip on the unknowable? Here is Martin Rees wrestling with the "something rather than nothing" question: "The preeminent mystery is why anything exists at all. What breathes life into the equations, and actualized them in a real cosmos? Such questions lie beyond science, however: they are the province of philosophers and theologians."

Perhaps that is to give in too easily, but it is certainly true that science flourishes when we share the unknowable with other disciplines. If the unknowable has an impact on how we lead our lives, then it is worth having ways to probe the consequences of choosing an answer to an unknowable. Music, poetry, stories, and art are powerful tools for exploring the implications of the unknowable.

Take the question of whether the universe is infinite. There are interesting consequences if you believe that space goes on forever. The fact that across the universe there may be infinitely many copies of you reading this book might have a profound effect on the way you lead your life, even if you will never know whether it's true.

Chaos theory implies that not only my casino die but also humans are in some ways part of the unknowable. Although we are physical systems, no amount of data will help us completely predict human behavior. The humanities are the best language we have for understanding as much as we can about what it is to be human.

Studies into consciousness suggest boundaries beyond which we cannot go. Our internal worlds are potentially unknowable to others. But isn't that one of the reasons we write and read novels? It is the most effective way to give others access to that internal world.

What we cannot know creates the space for myth, for stories, for imagination, as much as for science. We may not know, but that doesn't stop us from creating stories, and these stories are crucial in providing the material for what one day might be known. Without stories, we wouldn't have any science at all.

Wittgenstein concluded his *Tractatus Logico-Philosophicus* with the famous line: "Whereof one cannot speak, thereof one must be silent." I think that is defeatist, as did Wittgenstein in later life. A better denouement would be: "Whereof we cannot know, there our imaginations can play." After all, it's by telling stories that we began our journey to know what we know.

That journey has always been driven by what we do not know. As Maxwell declared, "Thoroughly conscious ignorance is the prelude to every real advance in science." I certainly think that's true when it comes to mathematics. I need to believe that there is a solution, and that I can find it if I'm going to have any chance of maintaining my faith as I venture into the unknown. Being aware that we don't know is crucial to making progress. Stephen Hawking appreciates the danger of believing we know it all: "The greatest enemy of knowledge is not ignorance but the illusion of knowledge."

For me, the conjectures of mathematics, the things we haven't proved, are its lifeblood. It is the things I do not know that drive me to continue my mathematical quest. I want to know if the Riemann hypothesis is true, and whether the PORC conjecture to which I have dedicated the last few decades of my research is false. As Jacob Bronowski put it, "Human knowledge is personal and responsible, an unending adventure at the edge of uncertainty."

The importance of the unattained destination is illustrated by the strange reaction many mathematicians have when a great theorem is finally proved. Just as there is a sense of sadness when you finish a great novel, the closure of a mathematical quest can have its own sense of melancholy. I think we were enjoying the challenge of Fermat's equations so much that there was a sense of depression mixed with the elation that greeted Andrew Wiles's solution of this 350-year-old enigma.

It is important to recognize that we must live with uncertainty, with the unknown, the unknowable. Even if we eventually manage to produce a theory that describes the way the universe works, we will never know that there isn't another chapter in the story, waiting for us to discover it. As much as we may crave certainty, to do science we must always be prepared to move on from the stories we tell now. But that's why science is alive and will never ossify.

So maybe it is important that I embrace the uncertainty of my casino die as it rattles around in my hand. And once it falls from my palm, perhaps not knowing how it will land will drive me to keep looking.

ACKNOWLEDGMENTS

Many thanks to the people who helped to make this book possible:

My editor at Viking: Joy de Menil.

My editor at 4th Estate: Louise Haines.

My agents: Antony Topping and Zoë Pagnamenta.

My assistant editor: Sarah Thickett.

My illustrator: Joy Gosney.

My readers: Andreas Brandhuber, Joseph Conlon, Pedro Ferreira, Chris Lintott, Dan Segal, and Christiane Timmel.

My interviewees: Bob May, Melissa Franklin, John Polkinghorne, John Barrow, Roger Penrose, and Christof Koch.

My employers: The Mathematical Institute, The Department of Continuing Education, New College, the University of Oxford.

My patron: Charles Simonyi.

My family: Shani, Tomer, Magaly, and Ina.

FURTHER READING

Al-Khalili, Jim. *Quantum: A Guide for the Perplexed.* Weidenfeld & Nicolson, 2003.

Armstrong, Karen. *The Case for God.* Bodley Head, 2009.

———. *The Great Transformation: The World in the Time of Buddha, Socrates, Confucius and Jeremiah.* Atlantic Books, 2006.

———. *A Short History of Myth.* Canongate, 2005.

Ayer, A.J. *Language, Truth and Logic.* Victor Gollancz, 1936.

———. *The Problem of Knowledge.* Penguin Books, 1956.

Baggini, Julian. *Atheism: A Very Short Introduction.* Oxford University Press, 2003.

Baggott, Jim. *Farewell to Reality: How Fairytale Physics Betrays the Search for Scientific Truth.* Constable, 2013.

Barbour, Julian. *The End of Time: The Next Revolution in Physics.* Oxford University Press, 1999.

Barrow, John. *Impossibility: The Limits of Science and the Science of Limits.* Oxford University Press, 1998.

———. *The Constants of Nature.* Jonathan Cape, 2002.

Barrow-Green, June. *Poincaré and the Three-Body Problem.* American Mathematical Society, 1997.

Bayne, Tim. *The Unity of Consciousness.* Oxford University Press, 2010.

Blackburn, Simon. *Truth: A Guide for the Perplexed.* Allen Lane, 2005.

Blackmore, Susan. *Consciousness: An Introduction.* Hodder & Stoughton, 2003.

———. *Conversations on Consciousness.* Oxford University Press, 2005.

Bondi, Hermann. *Relativity and Common Sense: A New Approach to Einstein.* Doubleday, 1964.

Borges, Jorge Luis. *Labyrinths: Selected Stories and Other Writings.* New Directions, 1962.

Butterworth, Jon. *Smashing Physics: Inside the World's Biggest Experiment.* Headline, 2014.

Carroll, Sean. *From Eternity to Here: The Quest for the Ultimate Theory of Time.* Oneworld Publications, 2011.

Close, Frank. *The Infinity Puzzle: Quantum Field Theory and the Hunt for an Orderly Universe*. Oxford University Press, 2013.

———. *Particle Physics: A Very Short Introduction*. Oxford University Press, 2004.

Conlon, Joseph. *Why String Theory?* CRC Press, 2016.

Cox, Brian, and Jeff Forshaw. *Why Does E=mc²? (And Why Should We Care?)*. Da Capo Press, 2009.

Dawkins, Richard. *The Blind Watchmaker*. Longman, 1986.

———. *The God Delusion*. Bantam Press, 2006.

Dennett, Daniel. *Consciousness Explained*. Little, Brown, 1991.

Deutsch, David. *The Fabric of Reality*. Allen Lane, 1997.

Dixon, Thomas. *Science and Religion: A Very Short Introduction*. Oxford University Press, 2008.

du Sautoy, Marcus. *Finding Moonshine*. 4th Estate, 2008.

———. *The Music of the Primes*. 4th Estate, 2003.

———. *The Number Mysteries*. 4th Estate, 2010.

Edelman, Gerald, and Giulio Tononi. *A Universe of Consciousness: How Matter Becomes Imagination*. Basic Books, 2000.

Ferreira, Pedro. *The Perfect Theory: A Century of Geniuses and the Battle over General Relativity*. Little, Brown, 2014.

———. *The State of the Universe. A Primer in Modern Cosmology*. Weidenfeld & Nicolson, 2006.

Feynman, Richard. *The Feynman Lectures on Physics*. Addison-Wesley, 1964. Available at http://feynmanlectures.caltech.edu.

Gamow, George. *Mr. Tompkins in Paperback*. Cambridge University Press, 1965.

Gleick, James. *Chaos: The Amazing Science of the Unpredictable*. Heinemann, 1988.

Goldstein, Rebecca. *36 Arguments for the Existence of God*. Atlantic Books, 2010.

Greene, Brian. *The Elegant Universe: Superstrings, Hidden Dimensions, and the Quest for the Ultimate Theory*. Norton, 1999.

———. *The Fabric of the Cosmos*. Knopf, 2004.

———. *The Hidden Reality: Parallel Universes and the Deep Laws of the Cosmos*. Knopf, 2011.

Guth, Alan. *The Inflationary Universe: The Quest for a New Theory of Cosmic Origins*. Addison-Wesley, 1997.

Hawking, Stephen. *A Brief History of Time: From the Big Bang to Black Holes*. Bantam, 1988.

Kaku, Michio. *Hyperspace: A Scientific Odyssey Through the 10th Dimension*. Oxford University Press, 1994.

Kapitaniak, M., J. Strzalko, J. Grabski, and T. Kapitaniak. "The Three-Dimensional Dynamics of the Die Throw." *Chaos* 22, no. 4 (2012): 047504.

Koch, Christof. *Consciousness: Confessions of a Romantic Reductionist.* MIT Press, 2012.

———. *The Quest for Consciousness: A Neurobiological Approach.* Roberts & Company, 2004.

Krauss, Lawrence. *A Universe from Nothing: Why There Is Something Rather Than Nothing.* Free Press, 2012.

Kurzweil, Ray. *How to Create a Mind: The Secrets of Human Thought Revealed.* Viking, 2012.

———. *The Singularity Is Near: When Humans Transcend Biology.* Viking, 2005.

Lakatos, Imre. *Proofs and Refutations: The Logic of Mathematical Discovery.* Cambridge University Press, 1976.

Laskar, Jacques, and Mickael Gastineau. "Existence of Collisional Trajectories of Mercury, Mars and Venus with the Earth," *Nature* 459 (2009): 817–819.

Levin, Janna. *How the Universe Got Its Spots: Diary of a Finite Time in a Finite Space.* Princeton University Press, 2002.

Lightman, Alan. *Einstein's Dreams.* First Warner Books, 1994.

Livio, Mario. *The Accelerating Universe: Infinite Expansion, the Cosmological Constant and the Beauty of the Cosmos.* Wiley, 2000.

Maddox, John. *What Remains to be Discovered: Mapping the Secrets of the Universe, the Origins of Life and the Future of the Human Race.* Free Press, 1998.

May, Robert M. "Simple Mathematical Models with Very Complicated Dynamics," *Nature* 261 (1976): 459–467.

McCabe, Herbert. *God Still Matters.* Continuum Books, 2002.

Monk, Ray. *Ludwig Wittgenstein: The Duty of Genius.* Jonathan Cape, 1990.

Mulhall, Stephen. *The Great Riddle: Wittgenstein and Nonsense, Theology and Philosophy.* Oxford University Press, 2015.

———. *Wittgenstein's Private Language: Grammar, Nonsense and Imagination in Philosophical Investigations.* Clarendon Press, 2006.

Nagel, Jennifer. *Knowledge: A Very Short Introduction.* Oxford University Press, 2014.

Penrose, Roger. *Cycles of Time: An Extraordinary New View of the Universe.* Bodley Head, 2010.

———. *The Emperor's New Mind: Concerning Computers, Minds, and the Laws of Physics.* Oxford University Press, 1989.

———. *The Road to Reality: A Complete Guide to the Laws of the Universe.* Jonathan Cape, 2004.

Peterson, Ivars. *Newton's Clock: Chaos in the Solar System.* Freeman, 1993.

Poincaré, Henri. *Science and Method.* Thomas Nelson, 1914.

Polkinghorne, John. *Belief in God in an Age of Science.* Yale University Press, 1998.

———. *Quantum Physics and Theology: An Unexpected Kinship.* Society for Promoting Christian Knowledge, 2007.

———. *Quantum Theory: A Very Short Introduction.* Oxford University Press, 2002.

Ramachandran, V.S. *A Brief Tour of Human Consciousness: From Impostor Poodles to Purple Numbers.* Pi Press, 2004.

Randall, Lisa. *Knocking on Heaven's Door: How Physics and Scientific Thinking Illuminate the Universe and the Modern World.* Bodley Head, 2011.

Rees, Martin. *From Here to Infinity: Scientific Horizons.* Profile Books, 2011.

———. *Just Six Numbers: The Deep Forces that Shape the Universe.* Weidenfeld & Nicolson, 1999.

Saari, Donald, and Zhihong Xia. "Off to Infinity in Finite Time." *Notices of the American Mathematical Society* 42, no. 5 (1980): 538–546.

Sacks, Jonathan. *The Great Partnership: God, Science and the Search for Meaning.* Hodder & Stoughton, 2011.

Sample, Ian. *Massive: The Hunt for the God Particle.* Virgin Books, 2010.

Seung, Sebastian. *Connectome: How the Brain's Wiring Makes Us Who We Are.* Houghton Mifflin Harcourt, 2012.

Silk, Joseph. *The Infinite Cosmos: Questions from the Frontiers of Cosmology.* Oxford University Press, 2006.

Singh, Simon. *Big Bang: The Most Important Scientific Discovery of All Time and Why You Need to Know About It.* 4th Estate, 2004.

———. *Fermat's Last Theorem.* 4th Estate, 1997.

Smolin, Lee. *Time Reborn: From the Crisis of Physics to the Future of the Universe.* Allen Lane, 2013.

Steane, Andrew. *Faithful to Science: The Role of Science in Religion.* Oxford University Press, 2014.

———. *The Wonderful World of Relativity: A Precise Guide for the General Reader.* Oxford University Press, 2011.

Stewart, Ian. *Does God Play Dice? The New Mathematics of Chaos.* Basil Blackwell, 1989.

Stoppard, Tom. *Arcadia.* Faber & Faber, 1993.

Sudbery, Anthony. *Quantum Mechanics and the Particles of Nature: An Outline for Mathematicians.* Cambridge University Press, 1986.

Taleb, Nassim. *The Black Swan: The Impact of the Highly Improbable.* Allen Lane, 2007.

Tegmark, Max. *Our Mathematical Universe: My Quest for the Ultimate Nature of Reality.* Knopf, 2014.

———. "The Mathematical Universe." *Foundations of Physics* 38 (2008): 101–150.

Tononi, Giulio. "Consciousness as Integrated Information: A Provisional Manifesto." *Biological Bulletin* 215, no. 3 (2008): 216–242.

———. *Phi: A Voyage from the Brain to the Soul.* Pantheon, 2012.

Tononi, Giulio, and Olaf Sporns. "Measuring Information Integration." *BMC Neuroscience* 4 (2003): 31.

Watts, Fraser, and Christopher Knight, eds. *God and the Scientist: Exploring the Work of John Polkinghorne.* Ashgate Publishing Limited, 2012.

Weinberg, Steven. *Dreams of a Final Theory: The Search for the Fundamental Laws of Nature.* Hutchinson Radius, 1993.

Williamson, Timothy. *Knowledge and Its Limits.* Oxford University Press, 2000.

Woit, Peter. *Not Even Wrong: The Failure of String Theory and the Continuing Challenges to Unify the Laws of Physics.* Jonathan Cape, 2006.

Yourgrau, Palle. *A World Without Time: The Forgotten Legacy of Gödel and Einstein.* Basic Books, 2005.

Zee, Anthony. *Quantum Field Theory in a Nutshell.* Princeton University Press, 2003.

The following website for the *Stanford Encyclopedia of Philosophy* is full of great material: *http://plato.stanford.edu*

INDEX

abduction, 224

absolute time, 254, 297, 365

accelerating expansion of universe, 7, 9,
  177, 181–83, 205–16, 288, 365

acceleration, 261–65, 280

"aces," 112

Acta Mathematica (journal), 41–43

Adams, John Couch, 187–88

aging, 262–64

air, and elements, 79, 84

air resistance, 66, 276

algebra, 86–87, 374, 375–76

algorithms, 337–38, 366

Alhazen, 189

Alice's Adventures in Wonderland
  (Carroll), 153

Allen, Paul G., 347

Allen, Woody, 301

Allen Institute for Brain Science, 347–48

Alpha Centauri, 178

alpha particles, *96*, 96–97

alpha wave, 314–15, *315*

al-Sufi, Abd al-Rahman, 194

altitude sickness, 204

alzahr, 23

Alzheimer's disease, 313

Amiot, Lawrence, 277

amplitude, 129, 130, 133, 139–40

Analyst, The (Berkeley), 84

ancient Greece. *See* Greeks, ancient

Anderson, Carl, 99–100

Andromeda nebula, 194–95

anesthesia, 334, 335, 339, 345–46

angle trisection, 375

angular momentum, 282

animal consciousness, 317–19,
  322

Aniston, Jennifer, neuron, 324–27

ant and the band, 207–9, *208*

anthropic principle, 222

antimatter, 99–100

aphasia, 307

Apple computers, 322

Aquinas, Thomas, 295, 394–95,
  410

Arago, François, 188

Arcadia (Stoppard), 21, 54

archeological digs, 22–23

Archimedes, 83

Aristarchus of Samos, 179

Aristotle, 3, 4

  actual and potential infinity, 393–94

  brain and senses, 305

  categories of events, 23–24

  concept of matter, 79, 83–84, 92

  light and vision, 189

  rhetoric and, 368

Arithmetica (Diophantus), 372, 376

Armstrong, Karen, 172, 417

artificial brain and consciousness, 351–52

artificial intelligence, 277–78, 313, 425

Ascent of Man (Bronowski), 4

Asteroids (computer game), 196–97, 200–201

Astronomer Royal, 186, 187, 192

asymmetrical twins, 263–64, 280

atheism, 416–18

atomic algebra, 86–87

atomic clocks, 244, 263

atomic number, 87

atomic theory, 79–86, 88–95

atoms, 76–77, 79, 90–95

attractors, 62

Augustine of Hippo, Saint, 24, 241, 294,
  395

Avatar (movie), 330

axioms, 368, 380–81

ILLUSTRATION CREDITS

All illustrations created by Joy Gosney, apart from the following:

FIRST EDGE

Page

25: Dice pyramid © Raymond Turvey

44: Chaotic path. Constructed using "Restricted Three-Body Problem in a Plane," Wolfram Demonstrations Project: http://demonstrations.wolfram.com/Restricted-ThreeBodyProblemInAPlane/

61: Evolutionary fractal tree. Illustration adapted from images generated by the One Zoom Tree of Life Explorer: http://www.one zoom.org/index.htm.

67: Four graphs describing the behavior of the dice. Illustration adapted from M. Kapitaniak, J. Strzalko, J. Grabski, and T. Kapitaniak. "The three-dimensional dynamics of the die throw," *Chaos* ٢٢(٤), ٢٠١٢.

SECOND EDGE

Page

89: Atoms inside dice. Yikrazuul / Wikimedia Commons / Public Domain

91: Jean Baptiste Perrin's *Les Atomes*. J. B. Perrin (SVG drawing by MiraiWarren) / Public Domain

THIRD EDGE

Page

127: Reprinted graph with permission from the American Physical Society as follows: C.G. Shull, "Single-Slit Diffraction of Neutrons." *Physical Review*, pp. 179, 752. Copyright 1969 by the American Physical Society: http://dx.doi.org/10.1103/Phys-Rev.179.752

FIFTH EDGE

Page

283: Entropy. Illustration adapted from Roger Penrose's *The Emperor's New Mind: Concerning Computers, Minds, and the Laws of Physics*. Oxford University Press, 1989.

289: CCC diagram © Roger Penrose. *Cycles of Time: An Extraordinary New View of the Universe*. The Bodley Head, 2010.

SIXTH EDGE

Page

312: Neuron drawing. Reproduced with kind permission from Santiago Ramón y Cajal, Cajal Legacy, Instituto Cajal, Madrid.

332: Purity by Randall Munroe. xkcd.com: http://xkcd.com/435/

341: Wakefulness / Deep Sleep. Illustration based on images from Marcello Massimini, Fabio Ferrarelli, Reto Huber, Steve K. Esser, Harpreet Singh, Giulio Tononi, "Breakdown of Cortical Effective Connectivity During Sleep," *Science* 309, 2228–32: 2005.

344: Two images of 8-node networks. Reproduced with kind permission from the authors Giulio Tononi and Olaf Sporns, "Measuring Information integration," *BMC Neuroscience* 4, 2003.